EARTH DRAMAS

Ancient mysteries and modern controversies

PHILIP A ALLEN

To Carolyn

The painter's portrait and the physicist's explanation are both rooted in reality, but they have been changed by the painter or the physicist into something more subtly imagined than the photographic appearance of things.

Jacob Bronowski (p. 103), The Abacus and the Rose, *Science and Human Values*, 1972, Harper & Row.

I grew up to be indifferent to the distinction between literature and science, which in my teens were simply two languages for experience that I learned together.

Jacob Bronowski in *World Authors 1950–1970* (1975) by J. Wakeman, pp. 221–223.

Unnoticed by the hawthorn hedge that lines the field,
When winter winds bring their piercing chill,
And by the vine-twisted oak when dying flames of day,
In golden shafts of light on woodland carpet lay.
The rising moon whose halo hides the ancient stars,
Makes silver shadows after fading twilight's hours,
While dew-covered ground sleeps dark, serene,
But science has it never seen.
The winding river does not know,
The direction science urges it to flow.

TABLE OF CONTENTS

ACKNOWLEDGMENTS

I am greatly indebted to a relatively wide circle of friends and relatives, who, being non-specialists, have acted as sounding boards for an early draft of this book. Their comments have ranged from the grammatical to the stylistic and typographical, but more importantly have helped me to judge what could be expected to be known by the lay (that is, non geological) reader. I am grateful especially to Carolyn Allen, John Allen, Paul and Tina Bailey, Peter Burgess, Frances Clements, Tom Kempton, Ian Salisbury and David Shepherd for their comments and suggestions.

PREFACE

Science is a temple built with human hands. It is not an apparatus that relentlessly cranks out answers from ice-cold observations swallowed like mouthfuls of krill into a whale. There is no linear arrow of scientific progress. Instead, scientists may disagree over the meaning of the same observations made at the same time in the same place. Good ideas may be stifled for decades by conservative resistance. Dodgy ideas may be broadly accepted based on the status of their originator. Some scientists may, in the face of new observations that were not predicted, treat their hypotheses as plastic. Repeated modifications may require increasingly fantastical bending of the hypothesis until it becomes a monster. Yet the scientist clings on to it as a capsized sailor to driftwood in the ocean. Are these rare occurrences, isolated aberrations in an otherwise well-oiled machine, or are they typical of the scientific process? Disturbingly, they are more common than at first thought.

Is the scientific enterprise a sociological battleground of egos, vested interests, corporate 'group think', political correctness and prejudices of every complexion? If this is the trajectory of science, it is no surprise that it is prone to the occasional journey down a 'road from foolishness to fraud'[1]. At the very least, the undergrowth of science is a shady place where different interpretations are fermented and hide from the light of day. Scientists become actors in a drama with a storyline they did not themselves compose. The drama is not some inevitable part of a natural order, but is the creation of human thinking and interaction. If this is the case, why and how do certain positions win out over others, so that one progresses towards acceptance as 'truth' and the other is consigned to the trash?[2]

Fractured as it is, and allowing for the sociological machinations intrinsic to an enterprise carried out by humans, science has the potential to get closer to the truth about the natural world than anything else available at present. Yet for how long do we wait for the sociological noise to dissipate? Society cannot be expected to know the answer to this, because scientists don't either. To dispel the truthfulness of science in a swamp of postmodernism isn't going to help anyone. That is a helpless scepticism.

Science is not a hostage to the social factors that determine how the enterprise is carried out, but it is also not immune. As in other areas of human thought, science makes use of powerful world-pictures, which turn debates into dramas. These are far from neutral, cold and un-textured. They are emotional, symbolic and full of imagination, particularly in the geological sciences. Despite the fact that our world-pictures are strongly influenced by social, cultural and psychological factors, these competing world-pictures affect the process of scientific enquiry but do not strongly undermine the objectivity of a result that has been fought hard for, weighed, tested and replicated, and passed into the hallowed ground of acceptance.

The way that science is conducted by individuals and groups, and the way that it progresses as an enterprise of learning, are complex, interesting and sometimes surprising. As the American philosopher and scientist C.S. Pierce (1839-1914), sometimes referred to as the 'father of pragmatism', said:

'A man must be downright crazy to doubt that science has made many true discoveries'.

This view was more recently echoed by the biochemist Peter Medawar in *The Limits of Science*[3], who wrote

'Science is incomparably the most successful enterprise human beings have ever engaged upon',

but the path toward these discoveries that human beings are engaged upon is frequently tortuous and rocky.

The essays that follow give a flavour of some of the dramas that have arisen in the past and continue to strongly influence current controversies about the Earth and its environment.

Chapter 1 starts with a general overview of the way that dramas are made to connect an otherwise bewildering 'meteoric shower of facts'. It becomes clear that world-pictures are fundamental to the way that humans organize their thoughts about the world around them, and scientists are no exception.

Some controversies involve early insights that are stifled by conservative resistance borne out of their strangeness in relation to the prevailing world-picture, as in the case of Alfred Wegener and his views on continental drift discussed in Chapter 2. These good but radical new ideas have a habit of winning through eventually, but then often championed by another scientist who is viewed as less threatening or toxic. There is a sense of the wastefulness of this type of scientific progress, with original insight, rejection, stalemate, reinvention and final acceptance, taking place over periods of as much as a century. Acceptance or rejection of your views may be more about who you are and what you stand for, rather than what you are saying.

Chapter 3 covers a controversy that was dramatized as a war between science and religion, alleged to have started with a famous debate in Oxford in 1860, shortly after the publication of Charles Darwin's *Origin of Species*. Yet this was a phoney war that was retrospectively constructed as the implications of Darwin's work fanned out into the social arena. We see the development of an ultra-Darwinist world-picture with abrasive relationships with alternative ways of viewing the story of life and the affairs of Man.

Ideas sounding very much like science can in reality have various additions that take them into the realm of metaphysics. This is the case with the Gaia hypothesis introduced in Chapter 4. Gaian thinking at its core is almost indecipherable from a systems-type scientific analysis but encourages a kind of mystical ecology.

The following chapter concerns a new way of labelling the era in which Man's influences on the planet have been extensive and profound – the 'Anthropocene' or 'Age of Man'. We discover how controversies have arisen in the past about attaching labels or 'golden spikes' at important boundaries, and consider the significance of living in the Anthropocene for the way we view the embedding of humans in the natural environment. Yet the definition and understanding of the Anthropocene goes well beyond conventional scientific frontiers.

Some controversies arise from discussion of ideas that are not particularly tortuous or prolonged, but involve different ways of joining up essentially the same dots. The vivid debate of the global glaciation of the planet known as Snowball Earth is described in Chapter 6. The debate shows a familiar pattern – one of opposing world-pictures and the difficulty of finding a critical test of the hypothesis whose result will be binding on both sides of the argument.

Some big questions have taken 300 years to solve to a high level of confidence, like the age of the Earth discussed in Chapter 7. Starting with an initial stab at the age of the Earth from an analysis of religious writings, estimates have, in general, got closer and closer to the currently accepted figure of 4.54 billion years. Over time, each estimate made use of assumptions that proved to be incorrect, but as new discoveries spawned new technologies, a consensus was eventually reached. If scientific answers take this time to become a consensus backed up by watertight evidence, the question arises of at what stage is it safe for scientists to confidently engage with the publics, and in so doing whether the publics need to be tuned-in in some way to the provisional nature of scientific discovery and to its uncertainties without losing faith in it altogether.

The mass extinction of life 65 million years ago, which catapulted the dinosaurs into oblivion, is the subject of much rancorous disagreement at the present day. In examples such as this, there is no steady ascent towards a high plateau of consensus. Instead it is like a game of ping-pong, with each side of the argument making thrusts into the other's territory. Chapter 8 looks at both sides of the argument of an impact from an extraterrestrial body and the eruption of volcanoes as responsible for the extinction of the dinosaurs, concluding with questions about the nature of evidence. We discover a looseness in the language and meaning of the words 'observations', 'evidence' and 'facts'.

Chapter 9 covers a topic that originated from ancient legends found in a wide range of early cultures concerning a global flood. We track the influence of the Noah's Flood legend on scientific thinking and find evidence of catastrophic floods in geologically recent time that were rebutted by conservative thinkers. An ancient flood disaster is unraveled that shaped the identity of the Israelite nation in captivity in Egypt, and an Egyptian famine is attributed to the eruption of the Santorini volcano. Both legends from ancient Egypt are prone to 'smoking gun syndrome'.

It is claimed that there are examples in the history of geology and geophysics of 'zombie science', where it is very difficult to kill off a hypothesis, long after the hypothesis has suffered multiple failures in predictive tests. The idea of hot, narrow rising conduits (plumes) as part of a circulation in the deep plumbing system of the Earth is cited as an example, which we consider in Chapter 10. If the idea of plumes is zombie science, then the bulk of the scientific community that believes in this conventional wisdom is instead in the grip of a silent mass delusion. Irrespective of this particular case of controversy over the witch's cauldron of Earth's deep circulation, zombie science in general is particularly damaging since the translation to the publics of a conventional wisdom that turns out to be flawed undermines confidence in the pronouncements of scientists in general.

Chapter 11 deals with a new paradigm that took root without the normal scientific procedures of presentation of the supporting data, descriptions of methods and assumptions and testing by other scientists. It concerns how we describe and try to understand the architecture of rock bodies beneath the surface of the Earth. The 'sequence stratigraphy' paradigm was widely accepted by the scientific community on the basis of the high reputation of a multinational corporation where it originated. Part of the paradigm has advanced a methodology that has proven useful, while the other part concerning the construction of a curve of global sealevel variations in the past, and its use as a worldwide dating method, is in crisis.

The final topic concerns a subject where there is near-consensus across a very wide range of the science community, but where the translation of scientific understandings to the publics has been problematical. Chapter 12 concerns the carbon crisis in which we are currently situated. The basis for the concern over future climate change is addressed, focused on the use of fossil fuels, but the conflict imagery between environmentalists and oil corporations as saints and sinners is dismissed as unhelpful. The climate change debate illustrates the intervention of eco-fundamentalism into the translation of the scientific consensus to society.

The book concludes with some thoughts as to why scientific findings are commonly greeted with mistrust and cynicism in the publics, and proposes that science needs to humanized rather than imposed if we are to avoid dark sarcasm in the classroom.

1 MAKING DRAMAS: THE SECRET WORKINGS OF SCIENCE

In which we discover that science proceeds through dramas manufactured in a dark undergrowth of social factors . . .

Joining the Dots

The Huqf area of the Sultanate of Oman lies close to the Indian Ocean midway between its two major cities – Muscat in the north and Salalah in the south. Standing on a resistant ridge of sandstone that forms the edge of a canoe-like fold of rocks, and gazing westwards, one sees the flat, featureless plain of the Empty Quarter. On some occasions, very cold air descends from high in the atmosphere and blows violently across the Empty Quarter towards the Indian Ocean, bringing with it sand, silt and dust. On those rare nights, folding tables are blown over, camping lights are extinguished and evening anecdotes beside a flickering brushwood fire are deferred to another time. During the day, fieldwork is abandoned and thoughts turn to hot showers and cold beer. However, normally, the baking heat under the cloudless skies of the Huqf region is relieved as the sun falls below the horizon by a gentle, refreshing breeze from the ocean, which merely ripples shirtsleeves and rustles sleeping bags. A quietness descends, punctuated by the noises of pots and pans, a murmur of discussion, and occasional peals of laughter. Some have their laboratories with expensive mass spectrometers and high performance computers, but I have field boots shredded by contact with splintery limestone, a rucksac of samples and a beard full of dust.

Lying on a camp bed under the still, Arabian night sky, far from other sources of light, the stars are especially brilliant and seem to fill the heavens with illuminated dots. It is said that the patriarch Abraham was told by God in a vision that his descendants would be as numerous as the stars in the sky[1], a fitting metaphor for a number too large to count. It is difficult to look upwards into the Arabian night sky without at the same time looking inwards, feeling at once joyful and tearful, and wondering what it's all about.

Seeking to find some kind of order in the apparent chaos of this starlit sky, the mind tries to join up the dots into memorable shapes, which childhood memories told me were Orion the Hunter, the Plough, the Great Bear and so on. But the reality is that, without these childhood memories, there would be many ways to join up the illuminated dots to make different constellations.

The same is true of those children's puzzle books that require the reader to join up the dots to make the outline of a well-known cartoon character. Joining the dots is easy, even for a child, if they are numbered, but when the number sequence is stripped away, the exercise becomes less trivial. The problem is that everyone sees the same dots, but we do not know the constellation or cartoon character outline that we strive to find and are left with a multitude of ways to join them up. No two people will join the dots in the same way, except perhaps identical twins who have led perfectly identical lives. Being an identical twin, I should be in a good position to appreciate this, but my egg sharer departed to the Antipodes before the experiment could be carried out. We have suffered too many years of separation to make it possible now. The clock is never turned back.

The way in which we join the dots, or make meaning out of a blizzard of observations, is based on our world-picture, and world-pictures invariably involve emotion and symbolism. In fact, we create *dramas* to make sense of things, which are played out in the theatre of science as well as culture. We are bound to engage our imaginations when shaping our views of the way the world works. In other words, there are motives in everything, and yes, in science as well as in poetry. Science is not so pure, so impersonal and so coldly objective, that it is immune from the motives of those who practice it[2]. As Mary Midgley writes in *Evolution as a Religion* (p.2,

1985)[3],

> 'Facts are not gathered in a vacuum, but to fill gaps in a world-picture that already exists. And the shape of this world-picture – determining the matters allowed for it, the principles of selection, the possible range of emphases – depends deeply on the motives for forming it in the first place'.

With somewhat sinister undertones, the great Polish-Jewish-British thinker Jacob Bronowski (1908-1974) wrote:

> 'no science is immune to the infection of politics and the corruption of power'.

In other words, scientists bring to the table a set of 'prejudgments' (or 'forestructures')[4] that reflect their preconceived views of what to expect, and their expectations of what would prove to be clinching evidence of a result.

As long as it continues to be carried out by human beings, it will be impossible to sanitize science so as to be free of the influence of world-pictures. Indeed, the cogs of the great wheel of science turn because of disputation, which invariably is based on opposing world-pictures. The question therefore is not whether world-pictures are part of the scientific process. The challenge is to work out which world-pictures are in use and to bring them out into the open.

This is not to say that almost all scientists do not behave well, and are in a form of 'explicit social contract' [5] with each other, which channels their behaviour and allows them to trust each other. The point here is not whether scientists are trustworthy, but what world-pictures scientists devise in trying to understand Nature. Science, when it is practiced and used correctly, does not seek to gain power from its findings, but tries merely to understand. Or as Jacob Bronowski put it, 'Man masters Nature not by force but by understanding'[6].

Different Worlds

Our thoughts are largely guided by the attitudes we take, and these are translated into the views we have of the external world. Carrying certain attitudes in our breast pockets, we might make claims that are impossible to support by evidence, in which case the world-picture would be regarded as a bad one. For example, the eugenics movement was supported by a particularly competitive view of sociobiology that grew from Darwinism. The extent to which Charles Darwin was himself aware of the extrapolation of his views to the social arena isn't clear, but it was his cousin Francis Galton, who coined the term 'eugenics'[7], and Charles's own son Leonard became president of the Eugenics Society. Darwin himself viewed natural selection as 'natural improvements' and believed that given enough time evolutionary processes would result in a drift in the direction of enhancement, progress and improvement, however one defines them. We can only imagine what they talked about over dinner or by the fireside.

The eugenics movement was based on an extrapolation of Darwin's findings on evolution to human society, what might be called *social Darwinism*. It held that human attributes, such as intelligence, physical strength and moral character were due to genetic make-up and genetic inheritance, so that one race of Man may be considered objectively superior to another. Taken far from its original context, eugenics rapidly became the outworking of a highly pathological science. As in other forms of pathological science, the flaws in the theory supporting it were outweighed by the profound wish that it were true in the eyes of its adherents. Policies stemming from eugenics, such as compulsory sterilization for 'defectives', were introduced in many nations, from the USA to Scandinavia to the Soviet Union, but the brute force of institutionalized eugenics was felt most in the Third Reich of Germany, with the all too well known consequences. As Walter Gratzer[8] writes:

> 'The history of eugenics affords another measure of the degree to which opinions, which should be based on the sifting of hard-won scientific evidence, can in reality be moulded by political beliefs or social pressures, not to mention mere self-interest. Scientific principle under such constraints is apt to yield to the desire to be on the right side.'

Whether it be eugenics, or polywater[9], or other examples of scientific self-delusion, it is not the fact that scientists were wrong, which is how one learns to be less wrong next time, but that opinions originating from a powerful world-picture over-rode all other considerations.

Most science does not operate in this way and could not even remotely be described as pathological[10]. Scientists are not in the habit of telling each other lies, at least not frequently and intentionally. Nevertheless, scientific debate and controversy commonly arise not from the use of different data sets but by the use of different world-pictures. It is not as if data are jealously guarded from prying eyes. Scientists are generally so excited about their new data that they simply can't wait to spill the beans at the next conference, and to get the results published as soon as possible. This means that new data rapidly become assimilated into the scientific community, where they are scrutinized and tested. However, scientists are far less self-aware in understanding the world-pictures that affect the way that they make meaning out of hard-won evidence such as experimental results, field observations and computer simulations. It is as if there is something shameful in admitting that there may be a 'bias' involved, that they are perhaps not pure and simply shining the torch of scientific truth into the blackness of human ignorance.

The workings of the scientific method, and the history of science that results, are therefore influenced by the competing world-pictures of which scientists are themselves only dimly aware. This is a particular hazard in those scientific disciplines, such as the geological sciences, where many plausible explanations may satisfy the available data. These disciplines deal with problems that are frequently *underdetermined*[11]. It is no surprise that geological problems are underdetermined, since geology seeks to understand the narrative of Earth history from mere fragments of information, like piecing together an early history of the Jewish nation from the Dead Sea scrolls.

It is sometimes said, though less so these days, that physics is a 'hard' science, with the subliminal message that other sciences are somehow 'soft'. But geology is 'difficult' rather than 'hard'. Geologists often search for the 'explanation that works', which is sometimes called 'synthetic science'[12]. This, in my opinion, is not something that geology should be proud of, since explanations can *appear* to work, but may be entirely bogus. Nevertheless, in historical sciences like geology, the explanation that works may have a logic derived from the historical evidence of what happened, rather than having a 'proof' in the sense used in physics.

The Scientific Undergrowth

Science is not immune from the social, cultural and psychological factors that contribute strongly to certain beliefs or world-pictures. The stunning success of science in explaining the natural world over the last couple of centuries is evidence enough that science, world-pictures included, works. But this does not mean that it is linear, acting like an arrow flying towards a target called 'truth'. Far from it, the scientific enterprise is non-linear, complex and human, and scientists have different attitudes to evidence.

Things can go wrong – slightly wrong, in which case a flaw leads to a distraction that causes a delay in the finding of a satisfactory explanation, and badly wrong where, to use Robert Park's subtitle, the scientist goes on a 'road from foolishness to fraud'[13]. Even back in the nineteenth century, there were concerns about the way science was carried out, Charles Babbage[14] recognizing forging (the outright invention of data), trimming (the cosmetic massaging of data) and cooking (removing data that don't fit).

In the Geological Society's apartments in swanky Piccadilly in the heart of London, is a painting by John Cooke, dated 1915. Eight erudite looking gentlemen all have their heads turned towards some objects laid out on a table. One of the men has a white laboratory coat and carries measuring tweezers. Standing behind him on his left is Arthur Smith Woodward, the Keeper of Geology at the Natural History Museum, London, and Charles Dawson, solicitor and amateur archaeologist. On the wall above a largely hidden fireplace is a picture of an elderly Charles Darwin, which brings to the event a certain erudite and sombre significance.

The white-coated anatomist is busy measuring the dimensions of a skull. It is the skull of Piltdown Man, thought at the time to be the missing link between apes and humans. Its discovery caused quite a stir and influenced research on human evolution for decades following. The skull, together with other fragments including a jawbone with a couple of teeth, came from gravel beds at Piltdown in Sussex, a rural county in the southeast corner of the green, sceptered isle of England[15]. The bones were discovered by Charles Dawson, who immediately wrote to Arthur Smith Woodward concerning his findings. Smith Woodward reconstructed Piltdown Man by using the jawbone and skull found by Dawson. Although Smith Woodward thought that the specimens belonged to a female, the Piltdown find was discussed as a male – as the 'earliest Englishman'. Piltdown Man was thought to be a new species and to date from half a million to a million years ago.

In the years after the discovery in 1912, many other human fossils were found around the world, but none looked like Piltdown Man, with its large skull and ape-like jaw. In the following years, dating techniques improved, and in the 1940s the bones were dated using fluorine contents. Using this technique, the fossils appeared to be 50,000 years old, not half a million or more, which threw doubt on the ape-like features. Eventually, in 1953, Piltdown Man was officially announced to be a fraud, since the skull and jawbone came from 2 different species, the skull from a modern human, and the jawbone and teeth from an ape (an orangutan or chimpanzee).

At the time, anatomist David Waterson expressed doubt in 1913 that the remains came from a new species, and offered his view that the skull looked human, yet the jawbone looked ape-like. Likewise, continental archaeologists were consistent in their belief that the jawbone came from a fossil ape but that the skull was human. One even suggested that the jaw came from an orangutan and had filed-down teeth! However, all these warnings were not listened to, probably because those involved desperately wanted to have a British early human to rival the discoveries of Cro-Magnons and Neanderthals in France and Germany. It was a question of nationalistic pride causing collective blindness and self-delusion.

With a darker perspective, British palaeontologists in the early 20th century believed that increasing brainpower and therefore intelligence had driven human evolution toward what they regarded as the superior white European. To find remains that suggested that the brain developed first and that other anatomical changes followed gave support to this view of 'brain before brawn'. It was a false view, which became pathological through the deluded wishful thinking surrounding Piltdown Man.

Was this an honest mistake made in the excitement of the moment? Unfortunately, it was not. For the teeth had indeed been filed down to make them shorter and more human-like, and the bones had been artificially stained to make them look older. Piltdown Man was a spectacular science forgery, but who committed it? It has all the makings of a good Agatha Christie 'whodunnit'. Clearly, Charles Dawson had every opportunity and was the originator of the 'discovery'. After his death from septicaemia in 1916, no further bones were found in the Piltdown gravels. He has since been found to be guilty of many other archaeological frauds[16]. Smith Woodward also had the opportunity but his only motive could have been stardom - yet the fraud was amateurish and he was professional enough surely to realize that it would one day be exposed. There are other possibilities. One of them is the Jesuit priest Teilhard de Chardin[17], who had a reputation of being a practical joker and who helped Dawson at Piltdown. When it wasn't recognized as such, presumably he kept quiet about his prank. Another possibility, bizarrely, is Arthur Conan Doyle, the creator of the detective Sherlock Holmes. Conan Doyle lived nearby and knew Dawson. One can only wonder at his motives. Presumably, as the perpetrator, he believed that the discovery of the fraud would be 'quite elementary' to solve.

Whoever was responsible, the Piltdown Man affair is at the extreme end of the road from foolishness to fraud. At its heart are the effects of nationalistic hubris, to some extent gender prejudice, and sadly racism, which echoed broad social undercurrents in England at the beginning of the 20th century. There have been other frauds, but perhaps none as audacious. Scientific frauds can't go on for long without being discovered.

At a less extreme end of the spectrum than fraud and outright lunacy lies a broad scientific landscape characterized by political, social and corporate pressures and simple human frailty leading to error and self-delusion. There are warning signs or 'stigmata' by which flawed science can be spotted. The most common symptom or pathway must surely be for the scientist to identify so strongly with his/her hypothesis that he/she loses objectivity. After putting in all those long hours, having agonized for years on the solution to a problem, and having cherished the result of all this hard labour and intellectual investment, it is hard to accept the stinging criticism of others. Having devoted a career to advancing a particular viewpoint, to have one's contributions discredited is like being airbrushed out of history. It is too much for the ego to stand. So the hypothesis is shored up again and again in a desperate attempt to hold back the rising tide of opposition. The scientist can't let go and tries to assimilate new evidence in increasingly creative and ultimately unbelievable ways. He/she becomes increasingly isolated, bitter and outspoken as the ship finally sinks without trace.

When faced with making a humiliating climb-down, or fumbling about trying to find a reason to keep the faith, history shows that scientists invariably choose the latter.

This level of stubborn denial may seem peculiar, but such cases are littered through human history. In the first century of the Common Era (AD), the astronomer and geographer resident in Alexandria in the Roman province of Egypt, but who wrote in Greek, Ptolemy (ca. 90-168) proposed the existence of a great southern continent. This idea persisted for a millennium and a half. Successive voyages by brave explorers of European maritime nations found no evidence for such a continent - the Portuguese Bartholomew Diaz in 1487, his compatriots Vasco da Gama in 1498, and Ferdinand Magellan in 1519 - so its shorelines were pushed back, but its existence was never doubted. Ptolemy's scheme was amended but not scrapped. It was not until the voyage of the British explorer James Cook (1728-1779) in 1768, supported by the best navigational aids available at the time, that the idea of a southern continent finally succumbed to defeat in the face of evidence.

Irving M. Langmuir (1881-1957) was a physicist who had a passion for sticking his nose into other people's business, particularly when he could smell the scent of errant science. He identified a set of clues or stigmata[18] by which pathological science could be identified. Large effects are attributed to very small, barely detectable causes; effects are characterized by great statistical scatter; claims of great accuracy of measurement are made; fantastic theories contrary to normal experience are devised; criticisms are deflected by *ad hoc* excuses made on the spur of the moment. The result is that adherents rapidly grow, but support crashes like a burst bubble.

Robert Park has listed seven diagnostic signs that raise suspicions about the scientific method[19]. Their applicability depends on the nature of the piece of science or pseudoscience being conducted, but a few resonate in the field of geology. For example, discoverers may make their claims directly or at an early stage to the popular media, rather than to fellow scientists, which limits the self-correcting nature of scientific exchange. They may claim that a conspiracy has tried to suppress their discovery, or at least that they are leading a crusade against a conspiratorial external world in a David and Goliath type struggle. Anecdotal evidence is commonly used to back up the claim.

Ideas that seemed reasonable enough (or even exciting) at the time can be clung to as conventional wisdom long after they have become unproductive. Having repeatedly failed tests of their predictive power, they are forced to 'morph' creatively to deal with each new line of evidence that accumulates. Instead of being junked, they trade in anomalies, exceptions and special cases, and frequently appeal to esoteric arguments, but refuse to lie down and die. These ideas are zombie science. Caltech geophysicist Don Anderson[20] writes:

> 'Zombie science . . . . . cannot be falsified or killed, and instead exists in a nether world of dogma and superstition that lies between paradigm shifts. Contradicting facts are ignored and added as new, unpredicted properties or adjustments to earlier speculations. The underlying assumptions, which are no longer believed by anyone, are forgotten in the process.'

An example of zombie science in the geosciences is the idea of fixed continents attached to a rigid interior of the Earth, which continued to be believed long after evidence was found to support the idea of moving continents, the birth of new oceans and Earth's constantly changing geographies. The alternative view is of course of the mobile continents of plate tectonics. Another example, it is claimed[21], is that continental crust has been generated steadily for billions of years by separating out of the light elements after the initial gravitational clumping together of the early Earth from the debris of the solar system. This idea, which originated in the 19th century, took on increased vigour in the 1950s and 1960s, and has, according to the Canadian-American geologist Richard Lee Armstrong (1937-1991), become a persistent myth that is difficult to overturn. The alternative view is that continents originated by chemical differentiation of light elements only early on in the evolution of the Earth and that current growth rates are negligible.

We will see examples of warning signs from the scientific undergrowth[22] in the examples that follow, but we will also see examples of an establishment view suppressing genuinely good ideas. It is not all about scientists on the 'road from foolishness to fraud'. It is equally about a collective impermeability to new evidence that is striking in its strangeness and appears unsupported by an obvious mechanism. Torchbearers sometimes are proved to be essentially correct, but too late for them to know it before they go to their graves.

If I were a physicist, alas a mere dream, I would have a certain view of how good science should proceed. I would probably approvingly quote the sociobiologist E.O. Wilson (1929-)[23]:

> 'Science is the systematic enterprise of gathering knowledge about the world and organizing and condensing that knowledge into testable laws and theories.'

I would probably go on to amplify the point that credible science requires new results to be subject to testing by other, independent scientists so that they can attempt to reproduce those results. Credible scientists should modify or abandon their theories based on new evidence, rather than entrench themselves deeper and deeper in defense of their original idea. Critically for the physicist, experimental tests should be possible to be devised, and if you can't devise a test, then you are not doing science.

All of this makes sense, but doesn't quite work in geology, since in many cases the experiment has already happened and can't be repeated in exactly the same way. The experiment is part of Earth's narrative. Let's take the question of anthropogenic (man-made) climate change. Let's make the safe assumption that climate change is being driven principally by man-induced emissions of greenhouse gases superimposed on the natural variability of climate known to have taken place in the past. We want to predict the global climate in 100 years time. The climate system is so complex that we would need every relevant parameter in our climate model to be represented, for each one to be perfectly quantified and for the interactions and feedbacks to be correctly identified. We would have to be perfectly aware of tipping points and non-linearities. So what experiment can we devise? The Earth has already conducted this experiment or similar experiments in the past. Geologists know what happened when atmospheric carbon dioxide concentrations were very high in the past[24]. These past experiments don't enable the details of global climate to be predicted in 100 years time, but they do provide an intellectual framework within which climate science can be conducted to advantage. Oscar Wilde wrote (*The Critic as Artist*, 1881) 'The one duty we owe to history is to rewrite it'. The geologist might retort that history first needs to be written before we can rewrite it.

In geology, we cannot take the physicist's narrow view as to what comprises science. This means that the self-correcting mechanisms that keep physics on the straight and narrow are more tenuous. That makes geology subject to much philosophical wrangling, and such a rich area for retrospectively seeing the 'human' impact on the scientific process.

Turf Wars

One should be suspicious of conflict imagery, especially where there is no such conflict and instead a nuanced jostling for position. Conflict imagery has a better home in politics to enable the electorate to choose between two centrist parties that have slightly different ways of arranging the deck chairs on the Titanic.

Geologists and geophysicists have a long history of tension with each other, both in corporations and in academia[25]. These scientific disciplines or 'social worlds'[26] have contrasting ideas of how to go about their business, possess different skills and use different technologies. Going back two or three decades, geologists in general lacked quantitative and computational skills, and were reluctant to see their work in terms of posing an hypothesis that could be tested. They were often regarded, with some justification, as being bogged down in pointless data collection and to be shackled by an overly historical, inductive[27] approach. Geophysicists are therefore portrayed as not seeing the need for geology at all, and as regarding geology as a non-scientific, historical and therefore second-class exercise. On the other hand, geologists are portrayed as believing that geophysics is simply a tool using techniques and knowledge derived from physics to solve geological problems. There is a partially hidden reason for this banter. These different social worlds are themselves legitimated by having differences with other social worlds. Such legitimation means that their existence continues to have value. Someone else's ideology always appears more strident, muscular and even predatory than one's own.

Examples are given in the essays that follow where the outworking of the different social worlds of geology and geophysics can be discriminated. Ideas may be rejected outright, or kicked into the long grass, as a result.

Times have hopefully changed. Many geologists are now well trained in physics and play a vital role in the numerical modeling of Earth processes. Geologists have shown themselves to be adept at systems-type thinking, requiring advanced levels of integration or synthesis. The social worlds of geology and geophysics come together and interact in an increasingly large number of arenas. An example is the collaborative ventures of geologists and geophysicists in trying to understand the links and feedbacks between the deep Earth and its surface. Another arena involving geologists, geophysicists and geomorphologists is the study of the long-term evolution of landscapes. And would anyone doubt that a successful combination of geologists, geophysicists, geochemists and engineers is required to land the Curiosity rover on Mars and run it about the cratered surface for months or years?

The deep veneration of all things to do with physics probably originates from Descartes, who believed that physics tells us the clearest and fullest truth about the physical world around us. There is a view that if a discipline doesn't look like physics, then there is something wrong with it. Even social scientists, who should know better, have an inferiority complex towards physical sciences. However, in combining with geophysics, geology has the ability to provide a vital narrative approach instead of being hamstrung by an inbuilt, often reductionist belief that quantification solves everything. Far from geophysics being tarnished as a result, it gains explanatory power.

Clashes, Controversies and Publics

In the chapters that follow are examples of great debates on issues that have largely been solved but whose history of development reveals the nature of the scientific process under opposing world-pictures. Other examples are from much more contemporary issues where views may be polarized within the scientific community, so the 'jury is still out', or where there is scientific consensus but a reluctance of politicians and public to get on board.

The selected examples beg the question of whether there is a persistent flaw, or set of flaws, in the way in which scientists go about their work, and in particular in the translation of their findings to an often sceptical public. The public, or more correctly 'publics' since there is no such homogenous thing as a public, appears highly resistant to being 'educated', judging from their stated beliefs, despite our increasingly transparent world. This suggests that we are not looking at the problem realistically. The essential problem is not the gross ignorance of the publics in the face of the dropping of pearls of wisdom by scientists. Physicist Robert Park writes[28] that:

> 'It is not so much knowledge of science that the public needs as a scientific world-view – an understanding that we live in an orderly universe, governed by physical laws that cannot be circumvented'.

But this bleak world-picture of a universe governed by physical laws is unlikely to make much headway with the publics, chiefly because science cannot explain all[29]. Robert Park, however, takes the view (p.45, *op.cit.*) that:

> 'Of the multitude of problems that daily vex modern society, few it seems can be easily resolved without recourse to the knowledge of science'.

Are the children of war zones dying through a lack of application of science? Are natural disasters such as the typhoon of 2013 that hit the Philippines the result of a lack of scientific prediction? Is the gap between rich and poor caused by scientific illiteracy, or likewise for terrorism, genocide, grinding poverty, unemployment, or depression, hopelessness, or loneliness? No, 'scientific truth is exact but it is incomplete'[30].

To take an example used by theologian and chemist Alister McGrath in *Surprised by Meaning* (2011, p.41), if science has no methods for deciding what is ethical, yet ethical questions can't be dismissed as meaningless, then there is a limit beyond which science cannot go. It seems obvious, but perhaps isn't, that because science has limits, and has no methods for dealing with issues such as value and meaning, yet is carried out by we humans, then it is highly likely that the scientific enterprise will proceed unevenly and occasionally pathologically. Furthermore, it will find extreme difficulty in translating its findings to a society that operates primarily by non-scientific motivations. Society is not an empty and neutral vessel waiting to be filled with scientific insight. In short, science is in danger of talking to itself while society goes about its business of being human.

Instead, scientists need a new model. They need to find new ways that do not require the wholesale demolition of the social, cultural and psychological barriers to effective translation of their findings.

The climate change debate does not instill confidence that this is happening. The originators of information too easily fall into the traps of exaggeration and neglect of uncertainties, a lack of balance, and sometimes, but rarely, a distortion of the evidence that can only be called skullduggery. As for the publics, they instinctively treat an expression of uncertainty as 'not knowing'. We have become used to being told the answer. If you go to the doctor with a complaint, it is more reassuring to be given a diagnosis, even if it turns out to be wrong, than being told that it could be this or it could be that. So on the question of whether scientists should emphasize the uncertainties and doubts, they are damned if they do and damned if they don't.

In the UK, too, people don't like being told what to do. We are naturally a little anarchistic. Add to this the hint that the public are themselves to blame and you provoke defensiveness. Say that the planet is doomed, and you create fatalism. All of this is understandable and perhaps inevitable. However, what is less obvious is the use of different world-pictures held by all the players in these dramas – the scientists, the publics, policy makers and politicians, media personnel, captains of industry and executives of state-owned companies. By understanding these world-

15

pictures, we start to get a better impression of why environmental activists try to prevent Arctic drilling or block shale-gas projects, why a large number of Christian fundamentalists still believe in a young Earth created in 6 days, and why a high percentage of the inhabitants of Texas does not believe in man-made climate change.

Nobody is immune from having a world-picture, since belief is a basic attribute of the human brain, no matter how one might protest one's objectivity. There is a tendency to answer criticisms of our world-picture by appealing to the world-picture itself. For example, I recently heard on the radio someone saying that he had criticized astrology to a friend, who had replied 'Well, that's because, being a Sagitarrius, you are naturally sceptical'. This might make sense to the holder of the world-picture, but is hardly a convincing defense.

In the past world-pictures were largely religious in origin, simply because in the absence of any alternative sources of information, religious teachings were the only place to turn. The narrative of the Judaeo-Christian Bible was not simply a religious slant on the history of the world, but at that time *was* the abbreviated history of the world. There were no competing ideas. Religious understandings proved to be a poor source of answers to the burgeoning scientific agenda, but they were never intended to offer such information. This use of religious teachings as a source of answers for understanding how Nature works may continue to take place in some isolated cases, but world-pictures are equally likely to be religious only in appearance, involving collective, mutually supporting faith in some general goal, but in reality are secular, humanist or 'green'.

Geneticist Steve Jones wrote[31]:

'To the scientific method, faith is a vice; to believers, a virtue.'

Zoologist Richard Dawkins saw science as a faith-free zone. This is naïve. More than a century ago, T.H. Huxley was against the interference in science of dogmatism of any kind, whether religious or anti-religious[32]. For science can become a tool in the hands of the faith of atheism as well as a weapon against the faith of religion.

The impact or effectiveness of world-pictures, and the ability of scientists to translate their findings to society, are also related to a loss of trust in experts and their opinions. In the UK, there are plenty of reasons for this. In the 1980s, we had 'mad cow disease' and the sight of a government minister gleefully eating a beef burger to reassure the public that everything was OK. Weeks later, 4 million cattle carcasses were smouldering in vast heaps in the English countryside. The acrid smoke from these places of slaughter drifted across the English countryside and hung like a pall over the coffin of the beef industry. There were 176 cases in the UK of the disease being passed to humans as new variant Creutzfeld-Jakob disease. The European Union put an export ban on British beef in March 1996. The ban lasted for 10 years, despite the efforts of Prime Minister John Major to get the ban lifted as early as May 1996. This was a case of the public being falsely reassured. In contrast, we were warned that the influenza pandemic of 'bird flu' could risk billions of lives, and the *American Scientist* ran an article in 2003 with the alarming subtitle[33]:

> 'The world is teetering on the edge of a pandemic that could kill a large fraction of the human population'.

The public lost confidence in poultry products and steeled itself for the worst. The risk of bird flu is still with us, but we did not experience the scale of pandemic feared. The public started to wonder about the quality of the information they were receiving.

Moving to the political arena, British Prime Minister Tony Blair joined American military forces in overthrowing Saddam Hussein of Iraq in 2003. The invasion was on the basis of Iraq's alleged possession of weapons of mass destruction and for harboring terrorists belonging to the terrorist group Al-Qaeda. Both of these items of 'intelligence' turned out to be spectacularly false. The loss of trust of politicians by the general public deepened, and was evident in the vote by the British parliament to not join USA and France in taking part in a military deployment in Syria in 2013. This had the effect of stalling, perhaps permanently, any military action.

Collectively, the publics in the UK have become cynical and disinterested in expert opinion in general. There is dark sarcasm in the classroom. It is not surprising that despite the efforts of the media, politicians and their scientific advisors, websites, blogs and magazine articles, and international panels of experts, there is significant resistance to the dire warnings of future climate change. Such cynicism prevails as the environmental movement steps up a gear in its campaigns. Are the two linked, since the green lobby positively fosters a mistrust of scientists by the publics, and even more so of large corporations?

The foregoing discussion has a touch of the postmodern to it. Postmodernist philosophy holds that there is no objective truth since all of our truth claims are changing and culturally influenced. Consequently, science (or anything else) does not objectively describe the world around us. If this were true, the empirical method of science, which puts great value on evidence, would fall apart.

Science has not fallen apart, nor is there any danger of it doing so. However, the gap between research at the coalface of science and public interest, understanding and respect of science is in danger of changing from a fissure to a chasm. Developing an appreciation of how science works is not achieved by serving up glorious success stories of scientific discoveries made by faultlessly objective super-intelligent robots. The gap would be better bridged by society having a better understanding of the role of social, political and psychological factors in the scientific enterprise, just as scientists need to understand that these same factors operate in society and are not barriers to be torn down.

As Ernest Rutherford said[34],

'An alleged scientific discovery has no merit unless it can be explained to a barmaid'.

2 SHIFTING FOUNDATIONS: CONTINENTS ADRIFT

In which a ground-breaking idea meets a hostile reception for nationalistic and political reasons, and rises from the ashes as a new paradigm . . . .

Trespass of Air Space

It is sometimes felt by those who have suffered the irritation of having their ideas rejected by contemporaries that, given time, things will change and they will eventually be vindicated. This tends to produce a certain doggedness, a resilience that they will be proved right all along, and perhaps a bitterness that they are not understood or have been betrayed by the narrow-mindedness of the modern age. Sometimes this is true, as is the case with Alfred Wegener and his theory of continental drift. Wegener would no doubt agree with the German philosopher Schopenhauer (1788-1860) that all truth passes through three stages; first it is ridiculed, then violently opposed and finally accepted as self-evident.

In the case of Alfred Wegener (1880-1930), there were factors that made his hypothesis fail in his lifetime that were more to do with culture and politics than relating to the strangeness of bare scientific evidence and ideas. Geological evidence had been accumulating for decades, even in outline for centuries, before it was synthesized by Wegener. These observations and views did not appear out of the blue. However, a number of things conspired to provoke great hostility to his hypothesis. He was a meteorologist who had trespassed into the air space of geology and geophysics. In addition, he was a German shortly after the bitter experiences of the First World War. The hostility reached its peak in North America, which was moving towards pragmatism and democracy in politics and science. His grand theorizing, for this is how it was perceived, was a reminder of the lofty, authoritarianism of 'old' Europe. His observations, which were fundamentally correct and compelling, lacked a physical mechanism to coherently tie them together. Wegener died prematurely on his fourth expedition to Greenland unaware that his hypothesis would one day gain broad acceptance and that it would lay a foundation for the new paradigm of plate tectonics. The example of continental drift is therefore a vivid reminder of the role of cultural and political forces, one might even say prejudices, that fundamentally affect the course of scientific discovery.

Alfred Wegener was born in Berlin in 1880, the youngest of five children to a theologian and minister who at one time ran an orphanage. Two boys survived infancy, and they seem to have been a dynamic duo. The young Alfred was interested in all things scientific, and had difficulty in deciding what to specialize in. He studied physics, astronomy and meteorology at Berlin, Heidelberg and Innsbruck and took a doctorate in astronomy at Humboldt University, Berlin in 1905, but chose to move more seriously into the rapidly growing fields of meteorology and climatology. As an early indication of his prodigious enthusiasm and motivation, he set a new world record in 1906 with his brother for the length of balloon flight, and that same year participated in his first expedition to the hostile climes of Greenland. He set up a weather station and developed a technique of using kites and balloons to track air movements. He returned home in 1908 to take up a lectureship in the University of Marburg, his lectures shortly later being adapted into a textbook on meteorology, called *Thermodynamics of the Atmosphere*. In 1912, shortly before the outbreak of the Great War, he made public his first thoughts on what we now call 'continental drift' at a meeting of the Geological Society at the Senckenberg-Museum, Frankfurt.

By the outbreak of war in 1914, Wegener had carried out a second expedition to Greenland, married the daughter of a famous meteorologist, and settled back into a lectureship at Marburg. He was wounded twice, withdrawn from active service and worked in the army weather service for the remainder of the duration of the War. Incredibly, within a nation deeply immersed in the horrors of war, he managed to complete the first version of *The Origin of Continents and Oceans*, but not surprisingly at this time of great suffering, the book attracted little attention. There were other things to worry about than whether continents had broken free of their moorings and were drifting about like stricken oil tankers.

After the War, he and his family moved to Hamburg, where he worked for several years on a book *The Climates of the Geological Past*, and in 1922 published a further, fully revised version of *The Origin of Continents and Oceans* that truly kick-started the hypothesis of continental drift. It was heavily, even brutally, criticized, especially in the USA, and in his lifetime was never widely accepted. He died on his fourth expedition to Greenland in 1930, not knowing whether his hypothesis would ever be accepted but personally convinced of its essential truth.

Continents on the Move

So what was the struggle all about? The idea of mobile continents originates from the observation of the similarities in shape of the western coastline of Africa and the eastern coastline of South America[1]. In addition, the coastlines of Antarctica, India and Madagascar fit neatly into the eastern coastline of Africa. Furthermore, Wegener was aware of discoveries of ancient plants that showed intriguing similarities despite being now separated by thousands of kilometres of ocean. He also knew about the direction of former ice streams found in both South America and Africa that had left striated (scratched) bedrock and thicknesses of bouldery sediment from the debris carried by the ice. He proposed that at a time in the past, all the continents were grouped into one great landmass, which was dubbed 'Pangaea', meaning 'All-Lands', and had since dispersed.

There is no doubt that Wegener made a very serious effort to make use of a wide array of geological, botanical and climatic data, and knew the significance of what he was piecing together. He was certainly not superficial in his analysis, or cavalier in his conclusions. Admittedly, he had somewhat vague and unconvincing ideas about the mechanism for the movement of continents, but essentially left the matter open. This was to prove to be a fatal strategic error, but one that was unavoidable at the time.

21

It is important to understand that Wegener was not flailing around in a vacuum, trying to build an edifice on no foundations. Centuries earlier in 1620, the English philosopher, scientist and writer Francis Bacon (1561-1626) pondered on the meaning of the similarities of the opposing coastlines of South America and Africa. Much evidence had been accumulating for several decades about the changing geographies of the Earth's past before Wegener's intervention. The Prussian naturalist, geographer and explorer Alexander von Humboldt (1769-1859), who with his brother founded the university in Berlin that bears their name, for example, proposed that the continents bordering the Atlantic Ocean were once joined. The French geographer Antonio Snider-Pellegrini (1802-1885) published maps[2] in 1858 showing a reconstruction of a supercontinent for Late Carboniferous times (approximately 300 million years ago), which explained the similarities of plants growing in the swamps of the Coal Measures in Europe and North America. The fact that he interpreted the separation of the continents as due to the biblical flood is an illustration of the tension between biblical tradition and new observation.

Eduard Suess (1831-1914) was a leading figure in documenting these past geographical changes. He was an Austrian geologist born in London, who spent his childhood in Prague and then Vienna, where at the age of just 26, he became a professor of geology in the university.

Bearing in mind the location of his workplace, it is not surprising that Suess took an interest in the geography and geology of the Alps. From this vantage point, he came to appreciate a number of remarkable, interconnected notions. The first was that the rocks comprising the high peaks of the Alps were once deposited in a former deep ocean, which in 1893 he called 'Tethys', a sea goddess in Greek mythology, and a term in usage to the present day. He also concluded that to explain strong similarities of fossil plants and animals in now-widely separated continental landmasses, they must have been connected in the past by land bridges allowing faunal migrations. The connected system of continents joined into one was termed 'Gondwana' by Suess in 1861, after a region of central northern India containing fossils that could be used to make connections across the former supercontinent. In particular, he knew that the *Glossopteris* fern was found in South America, Africa and India. This plant group is about 250 million years old, being found in the rocks of the Permian period.

In his three-volume *Face of the Earth*, Suess put forward his views that in the past sea levels had risen and fallen, and that these sea level changes could be correlated across all the continental masses. Now, these observations are essentially correct and revolutionary in drawing attention to a highly dynamic Earth with a narrative of mountain belts emerging from disappearing oceans, but Suess didn't quite join the dots perfectly. He wrongly believed that a contracting Earth was responsible for these changes, and despite his insights into a dynamic Earth, could not make the final step to conclude that the continents had dispersed from a former aggregated state in a supercontinent – instead, he believed that the sea had flooded the once contiguous land areas of Gondwana. By inference, the flooded areas must therefore be underlain by continental material. Nobody at the time knew whether this was true or false.

Hitting the Buffers

Building on the ideas of Suess and others, Alfred Wegener clearly had tremendous insights into the way the Earth works. In his later editions of *The Origin of Continents and Oceans* he shows an excellent intuitive understanding of the spreading of the ocean crust away from mid-ocean ridges, and even realized that small ocean basins were geologically younger than older ones. With these insights, why was he treated as a heretic, particularly in the USA?

American geologists and geophysicists deeply despised the suggestions of Wegener. There may have been a number of reasons for this, but few of them concern the scientific reasonableness or soundness of the hypothesis. One reason centres on two end-member methods of understanding Nature. The first method involves the dreaming up of an idea or hypothesis and the directed, targeted collection of observations designed to test whether the idea is a good one or a bad one, which is called the deductive approach. The second method is the unbiased collection of large amounts of data from which general truths emerge or are synthesized, which is called the inductive method, with its great proponent Francis Bacon. A critic of the Wegener hypothesis, Henry Fielding Reid, wrote:

'There have been many attempts to deduce the characteristics of the Earth from a hypothesis, but they have all failed . . . This is another of the same type. Science has developed by the painstaking comparison of observations and by close induction, not by first guessing at the cause and then deducing the phenomena.'

There are many followers of this inductive approach to this day. But the collection of raw observations with no guiding principle of what they are going to be used for is the job of an idiot. Even Charles Darwin, that great amasser of and deliberator over observations, admitted to the futility of observation without the guidance of an idea. The British philosopher of science William Whewell (1794-1866) wrote[3]:

'The facts are known but they are insulated and unconnected . . . . The pearls are there but they will not hang together until someone provides the string.'

Jacob Bronowski wrote 'No scientific theory is a collection of facts', and that 'a jumble of disorderly and meaningless items' would not be of any interest to anyone[4]. American poet Edna St. Vincent Millay (1892-1950) spoke of 'a meteoric shower of facts' raining from the sky, that 'lie unquestioned, uncombined'[5]. Observations are confetti strewn like snow:

They lie as confetti strewn like snow,
While pealing bells fall on those below,
In slanting light shed from churchyard tombs,
And distant memories hide in shaded gloom.

The world behind a milky vapour veiled,
While musty smells from dying leaves exhaled,
Apples shine like coals among the withered leaf
And scattered paper pieces lie beneath our feet.

A cloud unquestioned, secrets yet untold,
Relic of a narrative, countless millennia old,
On graveled path damp from autumn's blast,
About a time, a place, a lost world past.

Induction and deduction are well-known terms describing the

scientific method, but they are perhaps end member abstractions of what really goes on. The American philosopher Charles Sanders Pierce (1839-1914) suggested the term 'abduction', which is known currently as 'inference to the best explanation'. When an observation is made, perhaps a new and surprising one, we then try to work out an intellectual framework wherein our observation makes sense. We then argue that that intellectual framework is true. The realization of an explanation by such a process as abduction might come as a flash of insight or it may dawn slowly and progressively. It may be unearthed from something stated long ago and forgotten, or it may be brand new. Solving a murder mystery, like Agatha Christie's 'Hercule Poirot' or Arthur Conan Doyle's 'Sherlock Holmes', has an aspect of abduction to it. It involves more insight than simply arriving at an 'explanation that works', since there could be, and probably are, many other explanations that work equally well, whereas the 'inference to the best explanation' requires rare insight into a way of explaining observations best.

Many 18th century geological thinkers had deductive tendencies and came up with a number of all-embracing, ideological theories of the Earth that for one reason or another spectacularly failed. Lord Kelvin (1824-1907), for example, estimated the age of the Earth from its rate of cooling over time from an initially molten state as 100 millions of years, then revised it down to 20 millions of years, but both estimates are a long way from the currently accepted figure of 4.54 billion years. There was therefore a movement towards collecting more data following the inductive method, which gained ground in the early 19th century, especially in Britain, but less so on the European continent, where a more authoritarian, top-down view continued to be held. In continental Europe, professors were viewed as almost as infallible as the Pope, and the situation continues to the present day in some notable institutions.

In the USA, geologists were faced with the huge challenge of mapping and documenting vast tracts of their largely unexplored continent. The American revolution left geologists suspicious of authority in general, and of European elites in particular. They sought for new ways of doing things rather than being tied by the apron strings to the Old World. They wanted a scientific method that was somehow democratic, egalitarian and pragmatic that allowed different viewpoints to be aired, and Wegener's grand and lofty theorizing fell into this background context like a lead balloon. As Ted Nield[6] suggests, characteristically tongue-in-cheek, Wegener needed a PR consultant.

It was not helped that Wegener's book was translated by J.G.A. Skerl, who no doubt contributed to its loftiness and who exaggerated Wegener's dogmatism. In the eyes of Wegener's American critics, he was arrogantly single-minded, not interested in other ideas, and an amateur in geology. He seemed little interested in the hard work of geological data collection and admitted to 'hasty' research in support of his grand idea. Wegener basically knew he was right and simply dug his heels in when his hypothesis was attacked with criticisms.

Wegener was known to be a meteorologist and atmospheric physicist, and there is the smell of a turf war in the hostility shown to him. After all, how could a meteorologist out-smart a community of geologists and geophysics when it came, not to balloons and jet streams, but to their very staple diet of rocks and maps and things?[7] Although it could be claimed[8] that Wegener was as competent a geophysicist as any, he was the wrong sort of geophysicist in their eyes. He also did not have a formal training in geology, and it was a geological problem at heart, notwithstanding the need to find a plausible mechanism. It should be remembered that the Americans criticized the hypothesis for its lack of hard observations, not simply because there were no mechanisms known at the time that might have been responsible.

In the mid-1920s, it was not a good time to be German. The Great War was still fresh in memories. Wegener had been wounded in combat in his nation's war on the 'free' world, information that was (perhaps foolishly) included in the Introduction to the third and fourth editions of his book. In essence, Wegener's continental drift hypothesis sank without trace, or almost.

Signs of Redemption

In Britain, things were less vehement. The theory was discussed at a meeting of the British Association in the coastal Yorkshire town of Hull in 1923, and there seems to have been a tacit acceptance that drift of continents might have taken place, but scepticism that it had taken place in the last couple of millions of years as Wegener had suggested. The 'fit' of the African and South American continents, and the similarities of tectonic structures, floras and faunas on either side do not seem to have presented too many difficulties. As for a mechanism, in Britain, geologists were familiar with the idea that continents adjusted their elevation by variation in their thickness, just as a large iceberg floats to a certain level in the ocean dependent on the thickness of the ice below the water line. This balancing act is called isostasy, and achieving a balance like an iceberg is known as Airy isostasy. Consequently, continents could be viewed as floating in a fluid substrate, so the material deep below the surface, in what geologists call the mantle, was considered to be rather weak. The idea that continents might be able to move relative to each other by sliding over this weak layer like a piece of toast slipping over a layer of jam, would not have been too surprising to those well versed in Airy isostasy. An alternative model, and one that was popular in the USA, was that the topography of the Earth reflected internal density differences, but did not require mountains to have iceberg-like roots. This idea did not permit relative movement of continental blocks since they were viewed as firmly fixed in place.

John Joly (1857-1933), professor at Trinity College Dublin, had a drift version of his own. He believed that heat released by radioactive decay would heat up the rocks beneath the insulating overlying continents, causing massive volcanic eruptions. The continents, envisaged Joly, were acting like an efficient blanket, causing the underlying Earth to heat up and melt. The same process was tentatively suggested as responsible in some way for continental drift. He was fundamentally right. He lyrically referred to the 'power of the infinitely little over the infinitely great', and of the 'unending flow of energy from unstable atoms wrecking the stability of the world'. However, the man who was most influential in developing the idea of radioactive heating as important in explaining the drift of the continents was not John Joly, but Arthur Holmes.

Holmes (1890-1965) is the author of one of the leading geological textbooks of all time, entitled *Principles of Physical Geology*, which was first published in 1944. I remember it well, since as a schoolboy studying geology, an abridged version of this book was one of the few that were carried from classroom to classroom and to and from home and school. The abridged version was light enough not to cause an incident with child

Health and Safety, but also enabled 17 or 18 year-old schoolchildren like me to share in the fascination of geology. The textbook was published in several editions. I still have a copy of the second edition (published 1965), completely revised, and reaching to 1288 pages, bowing down my bookshelf. It was bought on entering university in a mad flush of consumerism at Aberystwyth, Wales in 1971. Even after Holmes's death in the year of publication of the second edition, the book has continued to sell. We will return to Holmes's role in the continental drift controversy shortly, since he was a key player in supporting Wegener's hypothesis against extraordinary levels of resistance.

In the USA, there were some sympathizers of the drift theory, and one of them (Frank Taylor, 1860-1938) proposed his own form of continental drift in 1910. He presented his variant of the theory at a meeting in New York in 1926, which was especially convened to address this contentious issue, sponsored by the fledgling American Association of Petroleum Geologists. Taylor proposed that there were two former continental masses, situated in each of the Earth's polar regions, which drifted towards the equator, causing crumpling at their leading edges and tensional thinning at the trailing edges. Somewhat fantastically, Taylor believed that the movement of continents was caused by the gravitational attraction of the Moon, which he believed was captured by the Earth during the Cretaceous period, some 100 million years ago. As Holmes states in *Principles of Physical Geology* (p. 1199), apart from the improbability that the Earth was without a Moon prior to the Cretaceous, how would we explain older periods of mountain building, and how could such a mechanism avoid acting like a gigantic brake on the Earth's rotation?

Taylor's hypothesis was published 2 years before Wegener's, but escaped the passionate criticism of Wegener's theory. Whereas Taylor envisaged continental drift as caused by gravitational drag from the Moon, Wegener believed that it was caused by the elevation of an equatorial bulge. Both authors believed in a movement of continents from the poles towards the equator, which Wegener called *Polflucht*. There were therefore striking similarities in the two theories.

The hostility of Wegener's reception in the USA compared to that of Taylor, despite similarity of content, was perhaps partly because Taylor was an American, not a German. Wegener's theory was roundly condemned at the New York meeting. In fact the only person to defend it was the convenor of the meeting.

Some American geologists, rather than rounding on the fundamental idea of drift, asked the question of whether the geological evidence held water. In line with the finest inductive traditions, expeditions to Africa had to be veiled as neutral, rather than as a deductive test of the Wegener hypothesis. The comparison of rocks in the Karoo region of South Africa with their South American equivalents in Brazil particularly needed to be done. The obvious person to carry out this work was a South African geologist of Scottish descent with much field experience of mapping, named Alexander du Toit (1878-1948).

Du Toit was appointed geologist in the Geological Commission of the Cape of Good Hope in 1903, and started detailed and extensive mapping of South African geology, particularly of the Karoo region, which contained in its sedimentary rocks a record of an ancient glaciation. These rocks showed unambiguous evidence of the scouring action and dumping of poorly sorted debris of ancient glaciers nearly 300 million years ago. It was suspected that these distinctive glacial sediments were also found in South America. Making this connection would lend strong support to the idea that Africa and South America had once been joined. With his vast experience of South African geology, du Toit wanted to test all of its correspondences with the geology of South America.

It took Alexander du Toit 2 years to produce his results, which were published in 1927 as *A Geological Comparison of South America and South Africa*. He confirmed the transatlantic connections suggested by Suess and elaborated upon by Wegener, but his readers and funders largely dismissed his findings. The problem seems to one of the underdetermined nature of the problem. The observations made by du Toit certainly allowed for drift as Wegener proposed, but did not *compel* the acceptance of drift in the view of its critics. There is no doubt as to the personal support that du Toit had for Wegener's basic thesis, as can be seen from the title of his most influential work – *Our Wandering Continents: An Hypothesis of Continental Drifting* – published in 1937.

Geophysicists continued to demand a mechanism for mobile

continents and refused to accept that continental drift could happen. Yet Ted Nield points out (*Supercontinent*, p.147) that mechanisms were known since the early 19th century that might have been invoked. Firstly, as we have previously seen, Airy isostasy required the relatively buoyant continents to be floating on a weak underlying mantle. This rigid outer layer might conceivably detach or decouple at its base and move relative to the rocks beneath. Secondly, parts of the Earth's surface such as Scandinavia were known to be rebounding upwards following the last glaciation, demonstrating that continents were affected by a deep flow beneath their roots. The time scale of the rebound of the Scandinavian area was known to be rapid (~10,000 years), which gave an indication of the low viscosity of the mantle beneath the continents. The viscosity might be low, but if you were to bring a sample of mantle rock (called peridotite) to the surface and place it on a bench in a laboratory, it would be dark green and very solid. When buried to great depths and subjected to high temperatures, peridotite flows, but very sluggishly. Thirdly, the rocks making up Earth's mantle were known to be undergoing a circulation not unlike that of porridge in a saucepan heated from below. This process of convection was advocated forcefully by Arthur Holmes in the 1920s, to whom we now return.

Holmes made a quick start in his geological career, measuring the age of a rock using the radiometric method in 1913 when he was still an undergraduate at the Royal College of Science, now Imperial College London. His early work on radiometric methods allowed him to appreciate the discovery in the first few years of the 20th century that decay of radiogenic isotopes created heat[9], that rocks contained radioactive elements[10], and therefore that the interior of the Earth must be generating heat. He was able to apply these findings to the age of the Earth. He also realized that this internal heat production must mean that the deep Earth is circulating by convection.

In the mid-1950s a magnetometer was developed that allowed the magnetic field 'frozen' into rocks at their birth to be recognized. The British geophysicist Keith Runcorn (1922-1995) used this magnetometer to analyze rocks from India. The results showed that India had moved across lines of latitude for thousands of kilometres before hitting Asia as in a slow-motion car crash test. Runcorn seized on the idea of convection currents in the mantle proposed by Holmes, and rightly concluded that the Himalayas were due to the tectonic mayhem in the frontal collision zone between India and Asia.

The combination of the appreciation of convection deep in the Earth and the use of the magnetic field 'frozen' into rocks at the time of their formation was crucial evidence in support of continental drift. There was now at least an informal mechanism for the mobility of continents. The last chapter of *Principles of Physical Geology* contains Arthur Holmes's views on continental drift shortly before he died. It is a masterpiece of synthesis that pre-dates the emergence of Plate Tectonics.

A last gasp of hostility came from an unexpected quarter. The Russian geologist Vladimir Beloussov (1907-1990), who had a career investigating vertical Earth movements rather than large horizontal (tangential) movements, published an extraordinary attack on continental drift in 1962[11], referring to the 'total vacuousness and sterility of the hypothesis'. If it might be a posthumous source of consolation to Alfred Wegener, Beloussov felt the same way about sea floor spreading and plate tectonics. It was nothing personal.

Wegener's hypothesis was therefore roundly rejected by men who should have been aware of parallel developments in geology and geophysics, and who chose to ignore the very large quantity of geological (and occasionally geophysical) observations that required an explanation. Evidently, Wegener's hypothesis failed primarily because of a conflict of world-pictures. He and his detractors were not actors in the same geological drama.

The Platform for Plate Tectonics

Before the Great War, and shortly after presenting his 'heretical' ideas at the Frankfurt Geological Society, Wegener made an expedition to Greenland (1912-1913), and he returned after the war in 1920 and 1930, dying during the latter. Why did he go repeatedly to Greenland? He was desperate to make measurements of drift, and thought that Greenland and Europe were separating. He hoped to pick up differences in measurements of longitude by combining astronomical and triangulation methods. This he did, but the measurements indicated very rapid movement of tens of metres per year.

These measurements couldn't confirm continental drift as the errors and uncertainties were too large – but was this illusion a result of flukes, errors or the blinding effects of hope?

Wegener was not to know that the pursuit of geodetic proof was bound to be fruitless with the techniques at his disposal. He did not have satellites and GPS (Global Positioning System). But even if he had, he would have had to take repeated measurements over several years to clock the exceptionally slow drift of continental masses. By the time GPS was available, opposition to the theory had dissolved as a result of a new discovery coming from the field of marine geophysics – a quantitative understanding of the spreading of the ocean floor at mid-ocean ridges. This was geophysical evidence collected by geophysicists for geophysicists, and it resonated much more than the geological field evidence proffered by du Toit. The new geophysical evidence was no more convincing than the geological evidence layed out with increasing levels of sophistication by Suess, Wegener and du Toit, but it was quantitative and it was 'owned' by the new elite of geophysicists[12].

It is tempting to believe that there were double standards in the way in which geological evidence was dismissed as circumstantial, but geophysical evidence was not. At the time of debate of Wegener's hypothesis, geophysicists did not rush to the aid of fumbling, confused geologists with their superior viewpoints. Geological insights were dismissed that proved to be true in almost every way, then technical developments arose that gave geophysicists the ways and means to collect new evidence from the ocean floor, at which point the contentious geological hypothesis of continental drift became the triumph of geophysics known as plate tectonics. The breakthrough came from magnetic surveying carried out to detect enemy submarines in the ocean.

Deep beneath the ocean surface lies crust that bears a magnetization picked up from the prevailing magnetic field of the Earth during cooling of rocks from a molten state. Marine surveys of the magnetization of these ocean floor rocks revealed a striped pattern of normal and reversed polarity. It was known that the Earth's magnetic field had flipped in polarity from time to time, so the magnetic stripes indicated a progressive creation of new sea floor, which moved away from its site of

32

creation at mid-ocean ridges. This, then, was the solution to continental drift – the continents were pushed along like geriatric passengers in wheel chairs by the creation of new intervening ocean basins.

Critics of Wegener's hypothesis at the New York symposium in 1926 persistently pointed to the problem of the timing and rate of drift. Wegener had admittedly made the mistake of assuming that the drifting in the North Atlantic region had taken place in the last 1 or 2 million years, which implied very high velocities of movement. But his reconstruction for the Late Carboniferous, 300 millions of years ago, is perfect. Wegener did not have at his fingertips a history of continental movements for all of the Earth over all of geological time, but his hypothesis clearly showed the way to go.

It is now known that continental plates have continually been on the move, periodically assembling into supercontinents and periodically disintegrating into a pattern of dispersed landmasses. This happens with a repeat time of about 300 million years, so only three supercontinents (named Rodinia, Gondwana and Pangaea) are known over the last billion years.

Geologists such as John Tuzo Wilson (1908-1993) have demonstrated the repeated opening and closing of oceans, using the Atlantic Ocean as an example. The dispersal and reassembly of continental plates take place in two modes. In the first mode, termed introversion, plates disperse from a supercontinental core, opening up large ocean basins in their wake, and crumpling into mountain belts at their fronts, but eventually the continental fragments reassemble by a closing of the same oceans, rather like the participants in a Scottish folk dance. In the second mode, termed extroversion, the continental plates continue to move away from the continental core and travel outwards until they collide with other continental fragments on their journey away from the supercontinental core, as they must on a spherical planet.

All this was known as the new marine surveys were getting under way, so that the resulting geophysical discoveries fell like seeds into a fertile bed of geological understanding of Earth history, which then, as now, is by far the best way to discover the truth about Earth's story.

The forthright Ernest Rutherford said that 'All science is either physics or stamp collecting'[13]. In the case of the continental drift hypothesis, all that stamp collecting proved to be quite useful.

3 VICTORIAN MYTHS AND DIRGES ABOUT NATURE

In which a phoney warfare between science and religion is dramatized and evolutionary principles are extended to the affairs of Man . . . .

Conflict Imagery

There can be few developments in natural science that have had greater impact than the theory of evolution. Evolution is not a topic owned by ecologists or population biologists or geneticists or palaeontologists. The impact of the theory of evolution rippled across the wide oceans of human thought and human action. Despite the fact that many theologians on the one hand, and many scientists on the other, regard natural science as being essentially neutral towards religion, the theory of evolution is commonly seen as a battleground of conflict between science and religion. The dialogue between science and religion, initiated by giants such as Galileo and Newton, has been particularly active since the last decades of the 20th century[1].

The use that has been made of evolutionary theory since Charles Darwin (1809-1882) published his epic volume with the cumbersome title *The Origin of Species by Means of Natural Selection, or the Preservation of Favoured Races in the Struggle for Life* in 1859 shows the striking impact of world-pictures, particularly in the social arena, where a drama has been forged that has taken on a life of its own. The more we go back in time, the higher the likelihood of religious viewpoints being involved in world-pictures. But all is not necessarily as it seems, and it would be a mistake to simply regard religion and science as at war over evolution.

The historical backdrop for the late nineteenth century development of 'conflict' or 'warfare' imagery was the Enlightenment. The Enlightenment was the struggle for freedom from traditional structures and thoughts. The holder of such structures in western Europe happened to be the Roman Catholic Church. Natural science was seen as one of the intellectual means of achieving these freedoms, while the French Revolution was the icon of its political and social dimension. Darwin's ideas were gratefully accepted by late nineteenth century thinkers, with the imagery of 'natural selection' and 'the struggle for life' as a competition for supremacy on the battlefield of evolution. The agenda was social, cultural and economic, but the headline news was the conflict between science and religion.

The root of the conflict is a debate at a meeting of the British Association in Oxford on 30 June 1860. There seems little to support the distorted view that the debate was really the first time Christianity had ever been asked to square off against science in a public forum in the whole of its history'[2], not because it was not the first debate, but because, as we shall see, it was a debate about the evolution of living things that cut across questions of religious or secular world-pictures. It took place in the University Museum, a few steps from the Department of Earth Science and a few more from the multicoloured brickwork of the most massive piece of architectural Victoriana in Oxford, Butterfield's Keble College. The venue was very appropriate, as the University Museum is a Victorian monument to the splendour of Nature. Animals, birds and vines are hewn into the stonemasonry of every Gothic capital and arch - or at least, every capital until they ran out of cash. The Earth Science department has since moved to a concrete, steel and glass construction a few hundred metres away, which gives them a home that is much better suited to teaching and research, but in the process the department has distanced itself, and more than physically, from the stonemason's art, marble columns and wrought-iron filigree that speaks of the interconnectedness of Nature and science.

The debate set the tone of a perceived clash between science and religion, involving Bishop Samuel Wilberforce (1805-1873), who is portrayed as representing the religious viewpoint, and Thomas Huxley (1825-1895) representing the new scientific viewpoint. The debate is therefore popularized, as it happens incorrectly, as that of an ignorant cleric flannelling around desperately trying to undermine his opponent's position before finally being well and truly put in his place by the erudite Mr. Huxley. Less popularly known is that Wilberforce had written a careful review[3] of *Origin of Species* some weeks earlier, which he had summarized in his speech. Darwin himself was appreciative of the review, which he found 'uncommonly clever'. Wilberforce had identified weaknesses in Darwin's suggestions that caused him to rethink a number of issues[4]. Wilberforce was thoughtful and polite, starting the review with:

> '[Mr. Darwin's] scientific attainments, his insight and carefulness as an observer, blended with no scanty measure of imaginative sagacity, and his clear and lively style, make all his writings unusually attractive.'

The various parties went cheerfully off to dinner together after the debate[5]. There seems little doubt that Wilberforce and Huxley remained on good personal terms after the debate.

The popular version of the encounter between the two men appears to have developed decades later, having little to do with the reality of the debate in Oxford in the summer of 1860, and much more to do with the general currents of thought in England as the century drew to a close.

In addition, although the original dispute was about new thoughts on the origin of the diversity of life, it was not conducted along the lines of a disagreement between science and Christian orthodoxy at the time. This can be illustrated by the fact that many of Darwin's fiercest opponents were the scientific establishment of the time, and many of Darwin's closest supporters were the more liberal-minded of clergymen. That Wilberforce had no deep philosophical problem with Darwin's findings, contrary to the idea of a weak and waffling, out-of-touch clergyman desperately fighting off the clear and incisive facts of science, can be clinched by reading an extract from Wilberforce's own review of Darwin's *Origin of Species*[6]:

'We are too loyal pupils of inductive philosophy to start back from any conclusion by reason of its strangeness . . . . we have no sympathy with those who object to any facts or alleged facts in Nature, or to any inference logically deduced from them, because they believe them to contradict what it appears to them is taught by revelation.'

Here, 'revelation' refers to the teachings of the Judaeo-Christian Bible.

The debate in Oxford was hardly noticed, except in the form of a few passing references made in the newspapers. So what caused Huxley's successors to lead a crusade against Wilberforce that turned such a trifling affair into something approaching a holy war? Huxley was an important Victorian thinker who had a lot to say about the purpose and meaning of life, and strongly believed in the separation of scientific discovery and religious teachings[7]. This brought him into direct conflict with the clergy, who held a powerful position in thought about the natural order. Wilberforce was a bishop, part of the clerical establishment, whereas Huxley was not attached to the Church. Wilberforce was also a scientist; he was vice-President of the British Academy, and spokesman for Sir Richard Owen, the greatest anatomist of his day. The young, strident Huxley, also an anatomist, was in open opposition to Owen. Huxley spent much of his later life trying to establish 'professionalism' in science: that is, to move away from the 'amateur gentleman', what we might call the 'naturalist' (which ironically is precisely how Darwin described himself), to the specialized scientist. Huxley was also a self-made, self-educated man, brought up in a middle class family that had fallen upon hard times financially. From these beginnings his career was nothing short of stellar. He became Professor of Natural History at the Royal School of Mines in 1854, which is now part of Imperial College and the Department of Earth Science and Engineering, President of the Royal Society in 1883-1885, and President of the Geological Society from 1868-1870. To Huxley, Wilberforce was therefore a pampered and over-privileged amateur as well as being a cleric.

Samuel Wilberforce is perhaps better known as the son of William Wilberforce, who campaigned to end the slave trade. He shared his father's views and was active in the anti-slavery movement. Samuel, far from being a conservative stick-in-the-mud, was a keen advocate of liberalism. With a father so strongly engaged in social justice, it is not surprising that Samuel was deeply concerned about the ethical consequences of Darwin's work. In that sense, he captures the mood of many who now have the benefit of having seen a rampant form of social Darwinism in action. Pseudoscience masquerading as 'truth' can indeed lead to social harm. The cartoon image of Wilberforce during the 1860 debate in Oxford published by *Vanity Fair*[8] shows the Bishop wringing his hands, and no doubt gave rise to his nickname of Soapy Sam. This is amusing enough, but should not distract us from the prescience of his review of Darwin's *Origin*.

Modification by Descent

Very few scientists deny that organisms change in their characteristics over time; that is, offspring are modified by descent, and we can very loosely apply the term 'evolution' to this process. Such modification over time is well supported by fossil evidence of, say, the development of feathers in dinosaurs, or legs in vertebrates. In other words, if we arrange fossils in a chronological order of their occurrence, we can see progressive but step-wise changes in size and shape[9]. Although there may be debate about whether this chronicle of change is slow, smooth and gradual or essentially stationary but punctuated by rapid jumps, most people are agreed that change has taken place when viewed over long periods of time.

Understanding how change happens, however, is more complex. Darwin's contribution, and independently that of his contemporary Alfred Wallace, was to suggest that some individuals in a species are better suited or 'fitted' to their environment than others, and that statistically the fittest are more likely to survive to produce offspring than those less well adapted to their environment. In this way, giraffes get longer necks, tigers get longer and sharper teeth, and butterflies change in colour to be camouflaged. The best-suited individuals then pass on their genetic code to their offspring, who are born with a silver spoon in their mouths - already advantaged in the game of life[10].

Darwin called the process by which evolution took place 'natural selection', but this does not imply intelligent choice by another agency - Nature does the selecting. What Darwin meant was that organisms that were better suited to their environment, as judged by their ability to reach the age of reproduction, tended to be more successful, and so the character of a population of individuals would change. The philosopher of science John C. Greene[11] suggested that this process is better called 'differential reproduction', and the differentiating was a result of inheritance. Variations of potential value to successful reproduction were a result of random (that is, unpredictable) genetic mutation. Rather than being simply viewed as changes that happened at the time to confer advantage, Darwin repeatedly refers to them as 'improvements'. 'Natural selection' to Darwin was the same as 'natural improvement' and he used exactly this term in a letter to the geologist Charles Lyell, whose star was in the ascendant at the time.

Natural improvement was, to Darwin, inevitable. Although costly and slow, it was the great hope for mankind. Julian Huxley, the grandson of Thomas (who played such an important role in the beginning of this Victorian myth), with the benefit of hindsight of the functioning of genetics, referred to evolution being 'blind and mechanical' and 'efficient in its own way – at the expense of slowness and extreme cruelty'. To avoid such cruel works of Nature, Huxley the younger thought that man had to take control, to wrestle away from Nature the future direction of evolution. This was a stepping-stone leading to an 'evolutionary humanism', that is, to a sociobiological religion of the secular world.

We will return to the religion of sociobiology later, but before doing so we need to look a little more deeply into how evolution is thought to work.

Individuals within a species share the same gene pool, and modification by descent is thought to happen by genetic mutations that take place at the time of the copying of genetic information in strands of DNA from parent to offspring, which in mammals would be at the time of fertilization of an egg. Most such mutations are highly damaging, as for instance occurs through exposure to high levels of toxins or radioactivity[12], but a tiny fraction are good, giving added adaptation to the environment of the individual. Now imagine that the environment is changing or that other species are improving in competitiveness. Organisms that were formerly well adapted become poorly fitted for the new situation and may reduce in numbers over time because of a deficit in reproduction; eventually the entire population may become extinct. Other organisms may be advantaged by the change in situation, thrive and increase in number. This ever-changing process is fundamentally responsible for the fossils of extinct species and modern descendants found in sedimentary rocks and loose sediments.

This is where most explanations of evolutionary theory end, but highly unsatisfactorily, because so far we have an explanation for how individuals in a species change, but not how one species turns into another. Let's be clear, *individuals do not evolve*. They may get taller or shorter, fatter or thinner, darker or lighter, develop long necks or short tails, even grow pointed noses and sticky-out ears, but these are changes within a population, and do not constitute evolution. Evolution only works through inheritance by reproduction acting in populations of individuals. Evolutionary biologists believe that species then evolve over successive generations as *local* populations change so that two, separated local populations become different. When these individuals are piled up in beds of fossils in sedimentary rocks, there appear to be distinct breaks between one ancestor species and another descendant species, something that would not be apparent when looking at the variation of living things in the time frame of human experience.

If evolution requires local populations to change so that eventually they are unable to successfully reproduce with other local populations, then a global population of a species that is in constant communication should not evolve. Consequently, Man may not be evolving in the era of globalization[13].

In some cases *groups* of organisms comprising many species share enough in common to suggest descent of one from the other, as in the case of some fishes that have bony fin supports that look almost identical to the early appendages of land-dwelling tetrapods (four-legged animals). This evolutionary transition or jump took place between 377 and 364 millions of years ago, and a fossil dated 370 millions of years ago, discovered in 1988, has so many similarities with both lobe-finned fish and amphibian-like tetrapods that its affinities are uncertain. It is tempting to refer to this example as the 'missing link', but strictly speaking, we are simply observing morphological characters that look transitional, without firm evidence of an evolutionary connection.

Thomas Huxley was in a good position to be able to judge the significance of the fossil record, and proposed in the late 1860s that birds were descended from certain dinosaurs. He was a specialist in the fossil *Archaeopteryx*, which was found in limestones 150 million years old in Germany in 1861. The fossil looked like a dinosaur but had feathers along its long bony tail, resembling a bird. Huxley was well acquainted with the fossil record and was much more confident than Darwin himself, who did not believe that the fossil record provided bullet proof support of his ideas for the evolution of species. Huxley was supremely confident of the meaning of the fossil record, writing:

> 'History has embalmed for us [as fossils] the speculations upon the origin of living beings.'

The vigour of Huxley's support of Darwin is shown by his calling himself 'Darwin's bulldog', though he was more convinced by the generalities of modification by descent than of the details of the mechanisms of natural selection, which he never quite embraced. Indeed, his ideas were decidedly fixist compared to the fluidity of Darwins'.

The Red Queen

The idea of evolution being powered by competition between species in a world of limited resources is emphasized in the Red Queen metaphor. In Lewis Carroll's (1832-1898) *Through the Looking Glass*, published in 1871, the girl heroine Alice enters another world by stepping through a mirror. It is a reflected version of her own familiar world. The landscape is laid out like a giant chessboard and the chess pieces are human-sized characters. One of them, the Red Queen, is able to run at high speeds, since she is the most mobile of pieces on the chessboard. The book is the story of Alice's progress across this landscape, ending in checkmate of the Red King.

Alice meets the Red Queen, who periodically repeats 'Faster! Faster!'. She takes Alice by the hand and they run as fast as Alice can manage, yet

> '. . . . the trees and the other things around them never changed their places at all: however fast they went they never seemed to pass anything.'

After another breathless sprint, Alice is propped up by a tree and exclaims,

> 'Why, I do believe we've been under this tree the whole time! Everything's just as it was! . . . . In our country, you'd generally get to somewhere else – if you ran very fast for a long time as we've been doing'.

The Red Queen replies:

> 'Now here, you see, it takes all the running you can do, to keep in the same place.'

Using this book character, the Red Queen hypothesis was proposed as a principle governing evolution[14]:

> 'for an evolutionary system, continuing development is needed just in order to maintain its fitness relative to the systems it is co-evolving with.'[15]

The principle implies intense competition. Every 'improvement' in a species (to use the terminology of Darwin to mean a change that confers a selective advantage) leads to a disadvantage for other species competing for the same resources. To maintain advantage, species run faster, or camouflage themselves, develop spines or poisons, grow sharper teeth, and so on. For species in competition with other species, therefore, it is necessary to run fast to stay still in the biological arms race[16]. A key question is whether this Red Queen process continues forever, locking species into a perpetual competitive struggle, or whether it operates only when the environment changes.

If a species is unable to respond to a new challenge, it faces extinction, whereas a successful adaptation confers survival. Such a successful adaptation may involve the forming of a new species. Speciation and extinction therefore go hand-in-hand as responses to new environmental challenges.

The strength of the Red Queen effect is dependent on how intensely a species competes for resources. This is likely to vary widely. Some species get on with their relatively unexciting, well-adapted lives with minimal interaction with other species, such as the shells living on the seabed. In contrast, some species have complicated strategies for survival, involving specialized ways of getting food and light and ensuring mobility. These are much more exciting species, like the ammonites and many mammals, but they are prone to extinction when stressed by limits in resources or physical conditions. The clear implication is that patterns of extinction, including those of mass extinctions, are dependent on biological competition in the context of changing physical environments, mediated by the particular conservative or specialized strategies of species. This has profound implications for how the fossil record of diversity (the net result of extinctions and speciations) is interpreted.

Herbert Spencer (1820-1903), who coined the phrase 'survival of the fittest' in 1864, extended Darwin's ideas to the social arena, suggesting that all of Nature was competitive so that only the strong survive over the weak, as in the Red Queen principle. Co-operation, love and altruism were seen as unreal and always to give way to self-interest. That the laws of a brutal Nature, 'red in tooth and claw', should be extended to the social arena was championed by the Austrian philosopher Friedrich Nietzsche (1844-1900), whose philosophy was unashamedly to release what is strongest in us, to enable us to rise to a higher plane. To reach this higher plane means that humans must break free from the chains by which the weak impede them. The most important of these chains, thought Nietzsche, was Christianity, with its distinction between good and evil, its emphasis on the goodness of the meek, the yielding, the pitying. To Nietzsche there was no good or evil, only good and bad, or if you like, good and bad specimens. A good man is one who is flourishing, potent, strong; a bad man is one who is diseased, impotent, feeble.

The fiercely nationalistic Ernst Haeckel (1834-1919) was to Germany what Herbert Spencer was to Britain, a prophet of social Darwinism. Haeckel believed that the State functioned like an organism, governed by the 'survival of the fittest', and was in an eternal struggle for existence. We are all too aware of how the thoughts of Haeckel and Nietzsche provided the driving force for Hitler's Germany.

The evolution debate provides an interesting angle on how conflict can be manufactured *post hoc*, and how principles can be taken far from their place of origin and used erroneously in dealing with moral, ethical or spiritual issues. This is not to deny that the greatest reason for the perceived conflict between science and Christianity undoubtedly emanates from the dogmatic stance of the collective body of individuals and beliefs that constitute organized religion. But that's another story.

Nature Red in Tooth and Claw?

If you were transported back to the fashionable coastal resorts of Victorian England, you might be surprised at the number of people picking around the cobbled stones and decaying cliffs, such was the interest in the new findings of geology. My copy of John Fowles's *The French Lieutenant's Woman*, first published in 1969, but set in 1867, is dog-eared beyond belief. After the book was turned into a film, I had great difficulty forgetting the vision of a young Meryl Streep standing on the harbour wall of The Cobb at the Dorset coastal town of Lyme Regis. Dressed in a black cloak, the wind whipping up her red hair as waves battered the foundations beneath her, she gazed sadly into the blackness of the sea. The lover of the melancholy lady in black was a geologist who spent his time collecting fossil sea urchins on the land-slipped cliffs of Jurassic topped with Cretaceous. He would have been familiar with the fossils entombed in those rocks, and how with every landslip a new treasure trove of ancient life was presented for discovery and collection. These assemblages of death strongly engaged the romantic minds of Victorian Britain, especially those of its poets. Shelley (in the lyrical drama *Prometheus Unbound*, 1820) speaks of the 'melancholy ruins of cancelled cycles'. Did Nature operate like a 'hammer of destruction'[17] in a coldly mechanical universe? Was Nature brutally indifferent to pain and suffering, indeed was it 'red in tooth and claw' as the English poet Tennyson wrote?

Seldom has a phrase been so degenerated and distorted by later use, implying a callousness of the living world and even a selfishness in evolution, as 'Nature, red in tooth and claw'. The savagery of Nature was in the premature taking of a young man's life, which is why Tennyson's poem is called *In Memoriam A.H.H.* – the initials of the man in question, Arthur Henry Hallam. Hallam and Tennyson met as undergraduates at Cambridge University. When Tennyson's first poems were harshly criticized by the reviewers of the day, AHH was there to sympathize. Hallam was engaged to be married to Tennyson's sister Emilia. Aged just 22, he suffered a massive stroke while travelling in Italy and died. Tennyson was devastated, and wrote his feelings in the poem *In Memoriam*, adding to it and editing it for 17 years before its eventual publication in 1850, just as Darwin was mulling over his *Origin of Species* and shortly after the publication of Charles Lyell's groundbreaking *Principles of Geology*.

Charles Smithson, the geologist of noble breeding who is attracted by the mystery of the French Lieutenant's Woman, talks lightly to his betrothed about a conversation he had the previous evening with her father, which echoes the 1860 debate in the University Museum in Oxford. He says,

> 'Your father ventured the opinion that Mr. Darwin should be exhibited in a cage in the zoological gardens. In the monkey-house. I tried to explain some of the scientific arguments behind the Darwinian position. I was unsuccessful.'

Later he continues the conversation,

> 'He did say that he would not let his daughter marry a man who considered his grandfather to be an ape. But I think on reflection he will recall that in my case it was a titled ape.'

The conversation highlights that the findings of scientists in the mid-1800s about the natural world were much discussed, and much disagreed upon. It was unsettling to think that Nature operated as a completely free agent, and shocking to believe that through natural processes well-heeled Victorian gentlemen were descended from the ape.

Victorian writers were accustomed to drawing from a wide range of art and new scientific discovery, in fact in a manner that makes modern 'sages' seem shallow, naïve and having knowledge without wisdom. Tennyson was aware of Charles Lyell's ideas on the long time spans of geological history and its slowly changing landscapes. This would have been disconcerting enough to the Victorian mind, but when extended to the biological realm through Darwin's findings, we have the combustible raw materials for a battle between political elites - the Church, and the emerging professional scientist. Tennyson therefore reflects in *In Memoriam A.H.H.*:

'There rolls the deep where grew the tree,
O Earth, what changes hast thou seen!
There where the long street roars, hath been
The stillness of the central sea.

The hills are shadows, and they flow
From form to form, and nothing stands;
They melt like mist, the solid lands,
Like clouds they shape themselves and go.'

We must return to Tennyson's grief. Here is the section (Canto LVI) of *In Memoriam* with the phrase 'Nature, red in tooth and claw'[18], and below it, my translation into modern lingo:

'So careful of the type?' but no.
From scarped cliff and quarried stone
She cries 'A thousand types are gone:
I care for nothing, all shall go'.

'Thou makest thine appeal to me;
I bring to life, I bring to death:
The Spirit does but mean the breath:
I know not more.' And he, shall he,

Man, her last work, who seem'd so fair,
Such splendid purpose in his eyes,
Who roll'd the psalm to wintry skies
Who built him fanes of fruitless prayer,

Who trusted God was love indeed
And love Creation's final law –
Tho' Nature, red in tooth and claw
With ravine, shriek'd against his creed –

Who loved, who suffered countless ills,
Who battled for the True, the Just,
Be blown about the desert dust,
Or seal'd within the iron hills?

No more? A monster then, a dream,
A discord. Dragons of the prime,
That tare each other in their slime,
Were mellow music match'd with him.

47

O life as futile, then, as frail!
O for thy voice to soothe and bless!
What hope of answer, or redress?
Behind the veil, behind the veil.

Nature seems as indifferent to the life of the individual as it does to the species found piled up in the fossil beds seen in quarries and cliffs. Nature responds to mankind, and the higher things that he can offer, such as a spiritual awareness, with extinction, as if he were no more than those fossilized remains. Is not Man's life therefore futile?

Man, whom God has made as the pinnacle of Creation, with such high ideals, and whose praises offered up to his God are left fruitless and unanswered.

Mankind thought that God was simply love, that it was programmed as part of the world He created. But the world brings violence, suffering and death. How can these be reconciled?

If, despite his good works and striving for justice, Mankind has nothing more to look forward to than extinction, then Nature is like a fictional monster in a bad dream. If so, frail and futile Man is at odds with the rest of Nature, and at odds with himself.

What hope is there of a way out of this contradiction, of this apparently futile life? The answer is hidden from view, too difficult to grasp. It is not for Man to know in his short mortal life.

With our limited life times, it is hard to appreciate that the average duration of a species is 4 million years. Yet this is short when set against the enormity of geological time. Some species have lasted as good as 500 million years by adopting a no-frills approach to life and survival, building a sensible body plan in harmony with the environment, and sticking to it doggedly and conservatively over time. Other species have been rather more exciting and experimental, but have commonly paid the ultimate price, that of extinction. One wonders if Man is of the latter variety. But extinction is not an optional extra or malicious after-thought; it makes way for speciation, without which all ecological space would be quickly filled.

Consider for a moment a world without individual death and without extinction of species. If two individuals have progeny, which themselves produce progeny, and so on down the generations, an initial population, which we can figuratively call Adam & Eve, grows to billions of individuals, even the soon-to-be-realized human population of the Earth of 8 billion, in a remarkably short period of time. The vacant ecological space of the planet therefore fills up amazingly quickly. Death and extinction are the only ways in which 'higher' life forms might develop. It is a *requirement* of a world that is poised with interesting evolutionary possibilities, including the emergence of Man, that death and extinction take place, not some bit of bad design by an accident-prone God.

In what sense then is Nature callous, or selfish, and in what sense is man's existence futile? This is surely a pessimistic distortion that Darwin himself would not recognize. It is a drama based on a certain world-picture, one dominated by the idea of Nature and society being individualistic and competitive from start to finish, of being hell-bent towards total destructive callousness. As the American sociobiologist M.T. Ghiselin wrote (1974)[19], 'Scratch an altruist and watch a hypocrite bleed'. Richard Dawkins's *The Selfish Gene* (published 1976) shares the same drama. Yet it is difficult to attribute selfishness and callousness to say, plants and microbes, which do not have motives. If selfishness is not meant at all, then what abstract causal property is envisaged? And if the whole world were selfish, then what would be the point of the word, since selfishness can only be appreciated by comparison with unselfish behaviour, which is denied to exist?

Let us not attribute this particular world-picture to Darwin. It is a construction of people far less nuanced and careful than he. As Mary Midgley wrote in *Evolution as a Religion* (1985), 'Nature is green long before she is red'. Darwin may have used the term 'struggle for existence' but thought of the process as more like mutual dependence. For every conflict, there is a backcloth of co-operation and social interaction. Indeed, Tennyson started with sadness and doubt, but finished his lament to his lost friend with hope. There is yet hope that the lion will lie down with the lamb.

Unfinished Business

Does Darwinism, with all of its modern additions, explain everything? Are we now just tinkering with the details? Is the path of evolution as random as the process of gene mutation implies? If we ran the tape of life on this planet again[20], would it produce a completely different outcome, or is there some inevitability in the emergence of animals with conscience like humans?[21] Can life navigate itself towards some solution to existence in the world (or worlds) it inhabits? And if it does, will it inevitably end up with animals with a conscience and a sense of moral responsibility? Or is the long pathway of life simply an accident and one of trillions of other viable possibilities, with the development of intelligence a very remote possibility? It would indeed be surprising if there were little remaining to find out about life on this planet. Surely, there is some evolutionary unfinished business?

The truly fundamental questions posed above circle around the topic of teleology – that is, the question of whether there is any direction or purpose in the natural world. Thomas Huxley, in reflecting in 1887 on the impact of Darwin's *Origin of Species*, believed that Darwin was neutral on the subject. It has been debated ever since. One view in evolutionary biology is that the development of life is entirely a matter of chance, that is, outcomes are purely random[22]. The evolutionary biologist Stephen Jay Gould (1941-2002) takes one end of the spectrum by insisting that everything in the history of life is essentially random and evolutionary processes are 'contingent', meaning that every step along the way is dependent on an uncertain previous occurrence. He writes (1989):

'We are the accidental result of an unplanned process.'[23]

Running the tape of life again, and again, and again, Stephen Jay Gould's view is that the cascade of life through time would each time be radically different.

However, another view is that evolution does not operate with contingency playing such a decisive role, and instead proceeds towards a limited number of end-points, so that re-running the tape always ends up with essentially the same result. This is termed evolutionary convergence, and is championed by the evolutionary palaeobiologist Simon Conway

Morris. Biological organization has a tendency, it is claimed, to arrive at the same destination in response to a particular need. It is like getting on a train anywhere in England, but always ending up in central London. Despite the innumerable genetic possibilities for the way a lineage of life may change over time, life's solution to a set number of needs, such as vision, or smell, is to converge on a limited number of successful outcomes. If this is so, it hints, or perhaps, shouts, at a deeper structure to life, played out over the vastness of geological time.

Let us make no mistake, the theory of evolution by natural selection is very widely accepted as the correct explanation for the diversity of life on Earth. Although some religious fundamentalist groups would not agree, it is clear that through the activities of science writers such as Stephen Jay Gould, Richard Dawkins, Daniel Dennett and Steve Jones, Darwinism has been brought forcibly before the general public. That organisms have changed over the immensity of geological time, and that some species have died out and been replaced by others is not in doubt. That experiments in evolution, seen by the remains of organisms found as fossils in sedimentary rocks, show a convergence towards certain recipes for evolutionary success, as in say the development of the eye, is also accepted. Yet, there is something disquieting about the self-confidence, the total conviction, the almost religious fervour of present-day ultra-Darwinists. The philosopher of science at Harvard University, John C. Greene[24], suggests:

> 'One would like to feel optimistic about the scientific mythology that has grown up around the theory of evolution, but it is hard to do so.'

Greene was very aware of the way in which evolutionary theory may take on features of myth. Ultra-Darwinists seem stridently sure of themselves and of the power in their own hands to chart the future course of Man. Contrast this with the humility of Isaac Newton, who marveled at his smallness in a vast, largely unknown universe:

> 'I seem to have only been like a small boy playing on the sea-shore, diverting myself in now and then finding a smoother pebble or a prettier shell than the ordinary, whilst the great ocean of truth lay all undiscovered before me'[25].'

There seems to be two ways of taking the word 'Darwinism'[26]. The first is 'Darwinism' as a widely accepted theory about the origin of species, which provides scientific understandings of the evolution of life that are continually being modified and deepened with further research. The second is 'Darwinism' as a world-picture, which offers a total and sufficient view of reality. This reality is thought to encompass the entire universe, against which rival world-pictures pale into insignificance[27]. This second meaning of Darwinism involves ideas that go far beyond scientific analysis, and takes on aspects of myth by the inclusion of metaphysical additions. It is an ultra-Darwinist world-picture.

There are therefore aspects of Darwinian thinking that go strictly beyond the scientific evidence and stray into areas that involve meaning and value, which have been smuggled somehow into the ultra-Darwinist project. John C. Greene argues that this panders to the often unspoken assumption in Western culture of a continuous and indefinite progress towards some elevated future state of man, who has climbed the staircase from alpha to omega. This in turn opens the door to the use of genetics to manipulate human nature into some ideal. Simon Conway Morris warns of this prospect, calling it 'genetic fundamentalism'. He worries[28] that some would treat the natural world as a sort of 'genetic play-dough'. Even if it were possible with any degree of success, who would determine the types of manipulation needed? Why, it will be the ultra-Darwinian geneticists of course. We had better beware.

4 GAIA: THE GREEN GODDESS

In which multidisciplinary science gains metaphysical additions and blends into deep ecology . . . .

Social Demands

How do we define what is science and what is not? Is science simply what scientists do in the course of their professional activities? If so, we have quickly given it a very wide remit, since the vast majority of scientists are applied scientists; that is, while carrying out their work to the highest theoretical or technical standards, they are fully aware of the social or political factors that loom large in carrying out their professions.

Abstractness of knowledge is sometimes claimed to confer a higher scientific status than the usefulness of knowledge of applied science[1]. Further, it is often claimed that applied science can only flourish by growing out of pure or basic science. Both of these claims tend to see pure and applied knowledge as separate entities.

Working in Imperial College London, and before that in ETH (the Swiss federal polytechnic) in Zürich, makes me all too aware of the way in which 'pure' and 'applied' research, rather than being separate entities, are like closely intertwined dance partners. Without the counterbalance of the other, the dance is likely to end with bodies lying on the floor. Many a 'pure' scientist has conducted his or her research only to find after a short time that their results have entered a field that is far from pure. The pathway of atomic physics to the bomb is perhaps the most graphic example. Without an understanding or self-awareness of these social and political factors, the pure scientist is likely to hit them in a destructive collision rather than through a coordinated dance.

A narrow view of science, as well as being prone to professional collisions, is a mechanism for disengaging from tricky problems, by saying they are 'nothing to do with me', and 'I can't be held responsible for the use made of my findings'. But I wonder if the same disengagement would take place if the tricky problem became an enterprise that could greatly benefit the human population of the Earth? Would we take the same standoffish attitude and abdicate all responsibility? I think not. In other words, the attitudes of the scientist are not immune from social demands. But is there a particular reason why science should throw its weight behind one project resulting from social demands rather than another? How do we rank the development of the bomb, the exploration for fossil fuels, genetically engineered crops, stem cell research and so on, and on what basis?

What is this future state of the world that scientists can play a pivotal role in helping to bring about? Or to be more precise, what kind of Superman[2] are we intending to create? Will these superhumans lord it over the proles, who are 'just manure in the soil in which are to grow the gorgeous flowers of elite culture'?[3]. Or is this an imaginary future for Man? Are there alternatives in which Man has a kind of minor walk-on part instead of being centre-stage?

So where does Man stand? One position is that the planet acts as a self-preserving system for life, but not necessarily human life, so that it takes on the properties of an organism itself. Another position, but not diametrically opposed, is that Man has affected the world's ecosystems so profoundly that we should name a new geological era in recognition of this. Both of these concepts, Gaia (this chapter) and the Anthropocene (next chapter), are loaded with the emotional, symbolic and faith-based paraphernalia that characterize world-pictures. The main protagonists are actors in their own dramas.

System Science with a Difference

Gaia, the name of a mother Goddess personifying the Earth in Greek mythology, has been coined as the name of an hypothesis that the Earth acts like an organism, which is self-regulating and directed to the goal of preserving life by its various feedbacks in a global system. Its originator, James Lovelock, in fact initially termed it the much less emotive 'Earth feedback hypothesis'[4] and intended it to explain the combinations of gases in the atmosphere that were fingerprints for the activity of planetary life. Lovelock had cut his teeth in the Jet Propulsion Laboratory in Pasadena, California, and was engaged in early thinking on how to recognize the activity of life on extraterrestrial planets and moons. A wise old bird probably tapped him on the shoulder one day and suggested that a title like this wasn't likely to lead to a gold rush to the bookshop, and was right. The 'Gaia hypothesis' sounds more interesting, but as we shall see, gave room for metaphysical, pseudo-spiritual and downright lunatic thinking on the fringe, and resulted in backward retrenchments after the idea was floated – a characteristic of big ideas that are pushed out with undue haste without the normal sifting and careful scrutiny found in the mainstream of science.

We can debate whether the Earth really does act as envisaged by proponents of the Gaia hypothesis, and in particular whether its pathway is directed to preserve life. But of immediate interest is what this hypothesis means in terms of the future of Man and the role of scientific discovery. Does the self-regulating Earth, which brought the dazzling forms of life out of the primordial ocean, approve of the humans that eventually resulted, together with the projects for human 'improvement' that scientists are engaged in? Will she pay our bills when we get things wrong, like a doting widow towards her reckless, spendthrift grandson?

This is the view of those, such as William Day, in *Genesis on Planet Earth, the Search for Life's Beginnings* (1984) who believe that we must play our part in the relentless onward and upward progress of Man towards some higher intellectual state of perfection, the 'Omega Point'[5]. Were the dropping of the atom bomb on Hiroshima, the concentration camps of Auschwitz and the racism of apartheid momentary lapses in human progress? There seems to be little to support this wildly anthropocentric view of 'bright-eyed superstition'[6] in the intentions of James Lovelock and the molecular biologist Lynne Margulis who co-launched the Gaia hypothesis[7]. Far from it, for Gaia is coldly indifferent to the success or failure of individual species, as is attested to by the merry-go-round of extinction and speciation over geological time. After all, 99% of known species are dead species, and geological history is littered with the debris of mass extinctions when half or more of all living things were removed from the face of the Earth forever.

No, Gaia looks after herself as a planet and is prepared to allow for some colateral damage in order to keep herself green. If that colateral damage involves the extinction of humans, so be it. Gaia is therefore an antidote to anthropomorphism, not a rich backer of human follies. To use the hypothesis as a means to boost the over-confidence in human progress shown by William Day is another example of the force of world-pictures in shaping scientific (or pseudo-scientific) thinking.

Gaia may therefore be regarded as a tool in the hands of those with a particular world-picture, but does it make sense? There are two areas where the original hypothesis seems to be particularly loaded with metaphysics; the idea of purpose and the idea of the Earth as a self-regulating organism.

Thus, Stephan Harding[8] writes:

'It is at least not impossible to regard the Earth's parts - soil, mountains, rivers, atmosphere etc. - as organs or parts of organs of a coordinated whole, each part with its definite function. And if we could see this whole, as a whole, through a great period of time, we might perceive not only organs with coordinated functions, but possibly also that process of consumption as replacement which in biology we call metabolism, or growth. In such case we would have all the visible attributes of a living thing, which we do not realize to be such because it is too big, and its life processes too slow.'

Perhaps even more problematically, is there purpose in Earth's trajectory? In what sense might a planet be said to be directed by purpose, in a kind of environmentalist's version of God. Is this not more the legitimate territory of religion? To answer that question, we need to first delve into what are the characteristics of faith-based philosophies and their adherents.

One of the most notable features of enterprises described as faiths is that they give their members a sense of being part of something bigger than themselves. Faith communities have broad goals that enclose those of the individuals that comprise them, and stimulate those individuals to work towards those goals, sometimes sacrificially, and sadly, sometimes brutally. In short, faiths are examples of the wholes being greater than the sum of its parts. Where faith-based movements work well, they are living realities of the value of joint belief and joint action, rather than superstitious clubs. Where faith-based movements work badly, they instill an uncritical 'group think' that can lead to prejudice and nastiness. Religious groups are simply faith-based communities where the faith is in some external single God, multiple gods or various spiritual forces. The positive characteristics of its members, among a range of negative, include a sense of awe, and a sense of transcendence or majesty, which brings the believer to worship and embraces the believer in a sense of goodness.

The Gaia movement, it seems to me, inspires many (but not all) of these same qualities. The Earth, as a self-preserving organism, is a whole greater than the sum of its parts. Mother Earth inspires awe and a sense of purpose, albeit sanitized of Man-centredness. It is an optional religion for atheists.

What then, does Gaia bring to the table that the integrated enterprise of Earth system science[9] does not? The key component of the Gaia hypothesis that sets it apart from Earth system science is the view that there is some active purpose in the interactions between the physical and chemical environments of the Earth and its co-evolving living world, that leads to the goal of maintaining optimal conditions for life. These interactions defend the planet from menacing external events, such as extraterrestrial bombardments, and internal events like massive volcanic eruptions. Some proponents believe in an 'intervention' by the co-evolving biosphere to maintain balance and to avert disaster, an interesting use of anthropocentric language.

An idea of how feedbacks take place is provided by the mathematical simulations, or 'parable', called Daisyworld[10]. The purpose of the Daisyworld parable is to show how organisms interact with their environment to produce feedbacks. It is used to demonstrate that the planet is self-regulating. It does this by simulating the energy balance of a planet subjected to steadily increasing solar radiation and populated by white or black daisies. The colour of the daisies controls the amount of incoming radiation that is reflected back into space (the albedo), and therefore affects the planet's surface temperature. As is commonly known, white objects reflect more than dark objects, so the white daisies cool the planet whereas black daisies warm it.

The black and white daisy populations compete in their growth: if there are too many black daisies the planet's surface gets too hot for optimal plant growth, and the reverse is true if there are too many white daisies. As a result the daisy populations adjust to produce a balance that favours a surface temperature that is close to that required for the optimal growth conditions of daisies. This is termed self-regulation. Lovelock and Watson hoped to show from the Daisyworld simulation that self-regulation can emerge completely automatically, without appealing to metaphysical concepts such as consciousness, purpose, pre-cognition or direction. However, leaving the metaphysics aside, whether the complex ecosystems that operate on the Earth, and have operated in its past, are easier to understand by reference to the Daisyworld parable is questionable[11].

Proponents of Daisyworld suggest that the white and black daisies compete according to Darwinian principles of natural selection. Yet black daisies don't change into new species of daisies, so the appeal to evolutionary theory appears gratuitous. It seems that the Daisyworld simulation simply shows that feedbacks may exist in the natural world. These may be positive feedbacks, which tend to amplify, or negative feedbacks, which tend to dampen or regulate.

A positive feedback can be illustrated by a casual look at a map of the ice cover of the world. Snow and ice are found in the Earth's polar regions where it is coldest. These surfaces are highly reflective. If the ice caps grow, the Earth becomes more reflective, causing a feedback towards greater cooling. This is the opposite of Gaian self-regulation.

Examples of the operation of feedbacks in Earth evolution that lead to equilibrium (the negative feedbacks that dampen wild oscillations), cited in favour of the Gaia hypothesis, are the long-term regulation of ocean salinity, atmospheric oxygen and global surface temperature. The

Gaia hypothesis holds that all of these forms of regulation are possible only through the intervention of biological processes. For example, if oxygen builds up to high levels of about 25%, about 4% higher than its present value of 20.95%, it is thought that the Earth's forests would be prone to wildfires, which would consume oxygen and lead to a reduction of oxygen production from photosynthesis. The presence of considerable amounts of charcoal in sedimentary rocks of the Carboniferous and Cretaceous periods due to burning, when atmospheric oxygen concentrations are thought to have been particularly high (above 25%) is supportive of this view.

Another example used is the role of biological activity in accelerating the drawdown of atmospheric carbon dioxide (carbon as a gas) during the chemical weathering of commonly occurring silicate rocks[12]. Chemical weathering is accelerated in warm climates characterized by high levels of biological activity. The chemical weathering releases dissolved carbon (carbon in solution in the form of the bicarbonate ion), which is transferred to the ocean by rivers, and combines with calcium to form calcium carbonate (carbon as a solid in the form of limestone), which comprises the skeletons of marine organisms. When these organisms die, they fall to the seabed and are buried beneath new layers of ocean sediment, thereby taking carbon out of the ocean-atmosphere system, driving global cooling. To a Gaian, this is an amazing example of self-regulation. To a geologist and oceanographer, it is simply the way the Earth works. No-one pretends that it is simple. No-one believes that life doesn't play a major role in the Earth's surface processes. And we only know about it through the work of scientists whom Gaians accuse of being reductionist, mechanistic and blinkered. This points then to the illogicality of Gaia. Gaians base their view of the interconnectedness of Nature on conventional scientific discovery, the object of their own criticism.

The idea that the Earth acts as a complex, integrated whole in which physical and biological processes are intertwined is not a particularly new idea. Nor is the related concept of the co-evolution of life and the physical environment[13]. Vladimir Vernadsky (1863-1945) was a Ukrainian mineralogist and geochemist who used the term 'biosphere', which had originally been suggested by Eduard Suess in 1885, meaning a realm of the physical environment primarily shaped by life processes. Vernadsky had met Suess in 1911. He deepened the idea of the biosphere to the 'noösphere' – the realm of human cognition. Vernadsky believed the emergence of a biosphere, and then a noösphere, were part of an evolutionary pathway of the Earth. Quite what this level of cognition is supposed to be is not clear. Is it a level of intellectual perfection obtained by a few, or a global sense of brotherly love and co-operation? To complete this pyramid is the idea of the Anthropocene[14], an age where every environment on Earth is permeated by the effects of Man, as we shall see in the following chapter.

The Gaia hypothesis therefore sits amongst these bedfellows. What makes the Gaia hypothesis distinctive is the various bits of clutter that are stuck to the theory, which on the one hand give the idea a mystical edge that has attracted fringe ideologies and pagan or New Age followers, but on the other hand has brought a torrent of criticism from the more reductionist of the bioscience and geochemistry communities.

Criticisms have been made on various fronts. One, made by the palaeontologist and philosopher Stephen Jay Gould, is that the hypothesis involves some pre-determined purpose, and was best seen as simply a metaphorical description of Earth processes. This is an accusation that sticks. Following the publication of the hypothesis in papers published in the early 1970s and in a popular book in 1979[15], Lovelock was at pains to point out in 1990 that

> 'Nowhere in our writings do we express the idea that planetary self-regulation is purposeful, or involves foresight or planning by the biota.'

Lynn Margulis[16] also stressed that Gaia was not a global 'organism', but

> '. . . an emergent property of interaction among organisms.'

However, the problem that arises is that if these aspects of the hypothesis are removed, then what is left that separates it from the conventional view of Earth processes? The Earth becomes a big interacting ecosystem in which the living world plays its full part.

Another criticism is that the Earth's systems do not, in fact, operate so as to achieve balance. As pointed out by James Kirchner[17], there is plenty of geological evidence that the Earth has undergone major excursions of chemical oceanography, climate and collapse of the biosphere that demonstrate feedbacks that are amplifying, not damping. The amplifying effect of the biosphere in the face of the burning of fossil fuels since the Industrial Revolution is an example. Surely, this is the whole point about the language of 'tipping points' in relation to global climate change?

The Gaia hypothesis has been modified since its original conception. It is worth noting that the hypothesis exploded onto the scene and was released in popular form at a very early stage in its development in the form of Lovelock's book *Gaia: A New Look at Life on Earth* in 1979. Lacking any significant early input from the scientific community, the idea has been forced to morph under waves of criticism. What is left is an interesting way to approach the closely interconnecting and inseparable worlds of biology, chemistry and physics, in which Earth system science is embedded. But had the Gaia hypothesis been first published in its revised, toned-down state, would it have grabbed much attention? It is likely that it would have smoothly joined the growing number of papers in bio-geophysiology pointing to the intricate yet powerful role of the web of life in Earth evolution, with which no-one fundamentally disagrees.

Eco-fundamentalism

What makes the Gaia hypothesis different to Earth system science are the add-ons, and these add-ons are essentially metaphysical. If one were to take the narrow definition of what comprises science, it is difficult to think of a test for the Gaia hypothesis. Since there is no test, then the Gaia hypothesis cannot be scientific. Yet the hypothesis has plenty of adherents, many of whom have 'green' credentials. So what is the relationship between the environmentalist lobby and Gaia supporters? The clue comes from the philosophy of 'deep' ecologists, or what might be called eco-fundamentalism.

I would describe myself as a pragmatic environmentalist, in the sense that I care about the hard evidence for what we are doing with the environment, wish to do what's necessary to protect it from damage, and believe that science and technology can help in this regard. I resist that dogmatism that sees organizations, industries, communities or individuals in terms of conflict imagery - as enemies, as immoral, as outside of the tribe.

Eco-fundamentalists base their world-picture on a particular type of ecology.

This type of ecology commonly involves some kind of mystical unity between mankind and Nature. That unity is viewed as harmed or even destroyed by science and technology. Owing to the recent overpowering growth of science and technology, Earth is thought to be in a state of crisis that will lead to an apocalyptic end. This set of ideas is not based on hard evidence but is more of a philosophy or faith that allows membership of the tribe. Eco-fundamentalists rarely publish their views in mainstream scientific journals, preferring instead in-house literature and a small number of periodicals such as *The Ecologist*.

Followers of 'deep ecology'[18] strive to attain (actually to re-attain) an 'organic' oneness between Man and Nature, and to correct for the very negative impacts of a rapacious, mechanical world dominated by science and technology. This is a perfectly legitimate standpoint to have, but it is a metaphysical position, not a scientific one. We can lament 'Leaving Eden'[19], but we have to be careful not to try to justify the raft of eco-fundamentalist policies as if they are derived from science. In fact, several warning signs flash up that eco-fundamentalism is the pathological science discussed in Chapter 1. This means that you can join an eco-fundamentalist organization with the good intention of saving the world, but you will not get a balanced assessment of the evidence to support the actions that you are subsidizing. Evidence very much plays second fiddle to the world-picture.

Prophets of doom from the eco-fundamentalist organizations have had a field day with anthropogenic climate change. Eco-warriors protest about greenhouse gas emissions, attempt to disrupt drilling in environmentally sensitive places such as the Arctic, and target operations to assess the potential of production of gas from shales using hydraulic fracturing. These protests may well come from perfectly good motives but they are not based on hard analysis of the long-term transition to renewable energy based economies. This message gets lost in the George Orwell-like mantra of 'renewables good, fossil fuels bad[20]'. Down on animal farm, the lights will go out.

The issue at hand is whether the Gaia hypothesis is linked to eco-fundamentalism or deep ecology, even simply by sharing a common metaphysical territory. The fact that you can take a course at Schumacher College (Totnes, Devon, UK) in 'From Gaia Theory to Deep Ecology'[21] proves the point. The director of the course, Stephan Harding, collaborates with James Lovelock in computer modeling of Daisyworld. He states:

'. . . we must see the Earth as an organic whole, with properties that are observable on the vast scale of the atmosphere and oceans. This has important implications for the way science is conducted and taught'.

But he then really breaks loose and let's his true feelings, for that is what they are, out in the open. He talks of the 'mechanistic, compartmentalizing conditions imposed on us since children by our society'. He talks of the operation of 'blind, meaningless laws of physics and chemistry'.

The idea of a machine-like world is not a new one that has arrived dressed in the green robes of Gaia. It perhaps started in 1665 when Isaac Newton concluded that the Moon was held in orbit around the Earth by gravity, and likewise the Earth around the Sun, since it turned the Earth into something like a clockwork mechanism. William Blake wrote in 1804 of 'dark Satanic mills'[22], meaning the mechanistic turning of the Industrial Revolution, seemingly without regard for life. Romantic poets felt the grinding bleakness of this view of the world acutely.

With Gaian thinking, there is an alternative to the 'blind, meaningless laws of physics and chemistry': there is

'a symphonic quality of this interconnectedness [of Nature], a quality that communicates an unspeakable magnificence'

that clearly echoes religious faiths. Harding goes on to say:

'You find yourself in a state of meditation, a state in which you lose your sense of separate identity.'

This sense of interconnectedness becomes the springboard for the opposition to what Harding calls 'all sorts of ecological abuses . . . , which is the hallmark of supporters of the Deep Ecology movement'.

Harding goes further, in believing that the dry, random, mechanistic, selfish, reductionist way of looking at the world is responsible for the massive social and environmental mistakes of Western-style development. Gaian perception 'helps to remedy this mental and spiritual plague'. A better way forward should be based on 'scientists' personal, deeply *subjective* (my italics) ecological experience'.

We live in an era of much greater environmental awareness than did our parents and grandparents. We live in an era of greater anxiety about the global and local environment. There is a growing sense of Man having screwed things up, that Nature is being constantly violated. The Christian value system, encouraging rampant consumption in a world of unbridled technological progress and human domination over Nature, has been seen by some[23] as the chief culprit. This view permeates the many modern writings of American theologian Matthew Fox and has given rise to the idea of creation-centred spirituality. Creation spirituality proposes that God is *in* Nature, not outside of it, which echoes the thoughts of Deep Ecology.

If anyone were in doubt as to the main driving force for the Gaia hypothesis, they should now know. It is fundamentally Deep Ecology dressed up as system science. It is a very particular, vivid world-picture of pseudoscience. World-pictures such as this have been given encouragement by the inroads of postmodern philosophy into contemporary thinking, to which we now briefly turn.

The Threat to Science of Postmodernism

Postmodernism is difficult to define. It holds that knowledge or claims of 'truth' are not what they seem, changing through historical periods, under the influence of different power structures and cultures. No single knowledge- or truth-claim is therefore more legitimate than another. Truth becomes merely a narrative undergoing continuous metamorphosis. If this is so, then the different possible definitions of postmodern philosophy must also be changing constantly, which is enough to make me junk the whole idea. Nevertheless, the reader will sense the threat that this poses for science, and in light of the preceding discussion, will have his or her awareness raised of the deeper philosophical currents in which Deep Ecology, eco-fundamentalism and the pseudoscience of Gaia are entrained.

A postmodernist view is that science does not objectively and truthfully describe the world around us. In other words, it is *relativist*. Any narrative is as good as another. If this were true, I doubt that scientists would have put Man on the Moon, eradicated smallpox, built the internet and put satellite navigation technology in our cars. However, we need to dig a little deeper.

I have some sympathy with the view that science is influenced by the particular moral, cultural and ideological context that the scientist is in. But postmodernists would go further. They would say that the *results* of the scientific process directly stem from these motivations, and if so, there is no objective truth to be had. If this is correct, then what is to stop anyone from promoting their own subjective view of the world, since evidence slides down the scale of importance? Indeed, the empirical approach of science, which puts great emphasis on observations and experiments, is devalued by this view. This is the framework upon which eco-fundamentalism, Deep Ecology and Gaia are built.

In an attempt to facilitate a more effective communication between science and society, the House of Lords Select Committee on Science and Society published a report called *See-Through Science*[24]. The title may simply be a plea for transparency, but seems to imply that science is somehow clandestinely corrupt and should not be trusted, rather like the cry for transparency to be shown by the banks, or governments, or multinationals. An extract from the report reads:

'Science is conducted and applied by individuals and as a collection of professions must have morality and values, and must be allowed and indeed expected to apply them to their work and its applications. By declaring openly the values that underpin their work, and by engaging with the values and attitudes of the public, they are far more likely to command public support'.

This statement looks innocent enough on the surface. Indeed, it looks refreshingly realistic about the social factors at work in science. However, it is dangerously vague in allowing the idea that the findings of science are never free of value judgments and therefore always lack objectivity. An important distinction has to be made between the activity of science, which is prone to social factors, and the hard-won findings of science, which are not. So let's be clear about how this approach is different to that taken in these pages.

I am suggesting that social, cultural and psychological factors affect our world-pictures, which in turn affects a number of processes. First of all, it affects the way we join up the dots. And second, it affects the way our ideas are received by the communities or publics into which the idea is placed. Dramas affect the process of science but given time and the rigour of disputation, the results of scientific enquiry are objective, testable and reproducible. If we intersect the scientific process at an immature stage, then we will certainly gain a misleading impression. Science and scientists aren't perfect, but the postmodernist idea that *all* of scientific discovery is subjective and unreliable is quite different. If this were the case, science would be useless, and the entire enterprise would disintegrate. The Lords on the Select Committee have got too much postmodernism about them for my liking. I doubt that many of them were well versed in the scientific method and with the history of science. But that may be part of my world-picture.

5 PARADISE LOST? THE AGE OF MAN

In which the pervasive impact of Man on Earth's physical and biological
systems is recognized, but should it be celebrated or lamented?

The Failed Masterpiece

When John Milton wrote (*Paradise Lost*, 1667) 'Accuse not Nature, she hath
done her part; Do thou but thine', I doubt that he quite foresaw what is
increasingly being called the Anthropocene[1], or 'The Age of Man'.

Nature has indeed done her part to bring the world to this point,
towards an increasing consciousness and self-determination of Man, as
foreseen by Vladimir Vernadsky (1863-1945) in his era or realm of human
cognition termed the noösphere, following the geologist turned philosopher
and Jesuit priest Teilhard de Chardin (1881-1955) who introduced the term
in 1922[2]. Rather than signifying the attainment by Man of some apex of
human thought and integration as Teilhard de Chardin imagined, we might
consider Man to still be much closer to alpha than to omega[3]. It might
instead be argued that the Age of Man brings with it unprecedented threats
of profound damage to this small blue planet that serves as our home, and
even the very extinction of our species. So much for our elevated level of
human consciousness on the onward and upward trajectory towards the
blissful state of Omega Man[4].

There is scarcely a vestige of our planet that has not succumbed in some way, subtle or dramatic, to the handiwork of Man. The coral reefs of the open ocean are in distress from rising water temperature and acidity, caused in turn by emissions of greenhouse gases through the activities of Man. There are so many dams on the world's waterways that natural or pristine rivers are a thing of the past. Many of the world's deltas, the dumping ground of conveyor belts of river-borne sediment, and the home of billions of people, are sinking beneath the waves. The oceans' fisheries are depleted, some of them to the point of exhaustion. Biodiversity is undergoing a big crunch, with unprecedented rates of extinction of species. And human population continues to go through the roof, from 3 billion to nearly 7 billion in the last half-century. So pervasive, so systemic, so global and yet so local are these impacts that many believe we should label our current time the Anthropocene. Some also believe that by labelling it so, we wake up to the reality of our responsibility for the future.

> For Nature has opened her eyes,
> And has for the first time seen herself,
> Through us poor humans,
> The failed masterpiece of creation.

The idea of a period of Earth history, albeit on-going, characterized by an increasingly important impact of Man on Earth systems comes originally from an Italian geologist called Antonio Stoppani (1824-1891), who combined being a geologist with being a Catholic priest and suggested the word 'Anthropozoic'[5] in 1873, meaning period of life dominated by Man. It's not clear why this didn't catch on, except that he was well ahead of his time. Judgments on the scientific value of an idea depend primarily on the environment into which the idea is introduced, rather than on the significance of the idea itself. There is an ideal time for ideas to be taken up, rather like warming up an audience for the main act. Evidently, the audience had not been sufficiently warmed up to receive Stoppani's ideas. It was left to the ecologist Eugene Stoermer and chemist Paul Crutzen to take up the torch in 2002, though 'Anthropozoic' turned to 'Anthropocene' in the process, a cosmetic change that fades into insignificance.

The story of this little known Italian geologist, Antonio Stoppani, is due for a reassessment in light of all the fuss about the Anthropocene. He was ordained as a priest in 1848 and in the same year manned the barricades of Milan in an attempt to secure Italian independence from the Austrian empire. His firebrand activities did not make him popular with the Catholic Church and he used his expulsion as an opportunity to study geology, becoming Professor of Geology at the University of Milano in 1867. He wrote provocatively about theology, Church, politics and the new nationhood of a liberated Italy, but he also delved into the impact on the world of the creation of Man, proclaiming the Anthropozoic era. Interestingly, Stoppani grasps the key ingredient in the Anthropozoic, namely that it is a unification of the 'physical' and the 'living' world, that

> 'a new and quite mysterious marriage unites physical nature to intellectual principle[6] .'

Stoppani marvels at the impact of Man in so short a time, compared to the aeons of deep geological time, writing 'how deep is Man's footprint on Earth already!'. But he clarifies what he thinks this footprint is – it is man's 'intellect', his 'intruding and powerful will'. It is striking that Stoppani believed that 'everything breathes with the strength of human intelligence', but he is writing in the late nineteenth century, before industrialization had despoiled air, river and ocean, and certainly before the carbon crisis and the dire pronouncements of the Intergovernmental Panel on Climate Change.

The ecologist Eugene Stoermer (1934-2012) used the term Anthropocene informally in the 1980s, and his lead was followed by others, such as Andrew Revkin[7]. Revkin, a reporter for the New York Times, has stressed that the Anthropocene marks not only the arrival of Man as a powerful, even dominant, agent for environmental change, but also that Man has himself become aware of this profound influence, like stepping up a rung on the ladder of cognition. The formal use of the term Anthropocene was made jointly by Stoermer and Crutzen[8], who received the Nobel Prize in 1995 for his work on the ozone hole. Crutzen was a signatory of the Humanist Manifesto, and an activist in matters from science education to nuclear winter and particularly global warming. His world-picture was therefore very different to Stoppani's, and Crutzen's emphasis is on the global impact of Man's activities, particularly in the atmosphere.

The Last Sacraments of a Disappearing World

My walk to the conference at the Geological Society of London on the topic of the Anthropocene (11 May 2011)[9] was not extraordinary in any kind of way. From the stucco-fronted classical squares of Kensington, with their neatly kept gardens, past the memorials to a diminutive queen, or rather to her German husband, along the margins of one of the world's great parks, where dog walkers chat and joggers with white earphones metronomically cruise, and mounted cavalry take shining horses to train in saw-dusted enclosures. There was nothing unusual about these scenes of peace and order. And there was nothing unusual about the dishevelled man in the subway with a little whistle that he played badly and intermittently, as if his musical attention span extended to a mere 4 bars. He stood beside a grimy sleeping bag that lay above flattened cardboard boxes, comprising a stratigraphy of homelessness. On the other side of the subway the world changed from green to urban-grey, save for a hotel on a corner with a veritable jungle growing up it, a Living Wall ten storeys high. The Living Wall screams at you that it is an anthropogenic biome, carefully designed and planted by Man, and watered and cared for by Man, and rather wonderful. It is a human ecology, a paradise on a street corner, not more than a quarter of a mile from the Geological Society, my destination this fresh day in early May.

Recognizing what you have done and owning up to it is a cathartic process that is necessary to go through if you want to make anything better, whether it's spilling red wine on your mother-in-law's new carpet or thoroughly messing up our disappearing world. But it is a good idea to separate the analysis of the impact of one's actions from the separate question of what we should do about it, because these are two different conversations, the first being scientific, the other involving a bit of everything, preferably including wisdom. Although these two conversations are linked, the first should inform the second. There are hidden dangers. If we get the scientific analysis wrong, we do not have the correct basis for making decisions about the way we manage the world in the future and perhaps carry around a flawed philosophical position guiding our actions. But the bigger danger is that the second conversation is hijacked by those with a range of pro- or anti-environmentalist agendas that they wish to advance, whatever the results of the scientific discourse turn out to be. In short, recognizing that we have entered the Age of Man does not provide a better moral, ethical or philosophical position to make important policy decisions on over-consumption, the futility and human waste of war, population growth, poverty and infant mortality. The moral compass for these issues is to be found elsewhere.

The Anthropocene is the recognition that natural ecological systems that have developed since the melting of the last continental ice masses from the northern hemisphere, a mere ten thousand years ago, have been so changed that they function differently. And the agent of change is Man. Delegates at the conference asked, 'When did this trend start?', 'What indicators do we use, what things do we measure?', 'What are our predictions for the future?', ' How long can we continue like this?', 'What will intelligent species in the future find when they look back at the beginning of the Age of Man?', 'Should Man and Nature be restored to some former state of innocence?'

If you can't say when a new geological period begins, and you haven't a clue when it will end, you are not on very solid ground for convincing others of its reality. Some would say that the Anthropocene begins with the Industrial Revolution of the late-eighteenth century[10], on the basis of the beginning of a major rise in the concentration of greenhouse gases, such as carbon dioxide, in the atmosphere. Others suggest that the Anthropocene started much earlier, coincident with the development of agriculture about 12,000 years ago. If we take this tack, there is a problem that we already have a term for the period of time since the melting of major ice masses (11,700 years ago), the Holocene, meaning 'entirely recent'. The Anthropocene must surely be recognized by the *pervasive* impact of the human population on Earth's ecosystems. But what do we measure? Options[11] are changes in physical sedimentation rates related to industrialization and agriculture; perturbations to the carbon cycle, recognized by rising atmospheric concentrations of carbon dioxide and methane, followed by global temperature changes; biotic changes recognized by animal and plant extinctions attributable to the impact of Man; and changes to the oceans in terms of global sealevel and water chemistry (especially acidification). Or if we follow the concept of the noösphere, it might be recognized by the achievement by humans of a higher level of consciousness or cognition, which might be the time at which humans first became aware of their pervasive influence. For some, this might be the time at which you read this chapter!

71

Whenever it began, and whatever it really means, the term 'Anthropocene' is a favoured currency in geoscience and its allied fields. A new journal has been launched[12], literature is awash with the Anthropocene buzzword, and the blogosphere is humming with opinion pieces. The International Union of Geological Sciences, which is not known for flippancy, has set up a committee to make a decision by 2016 on whether the Anthropocene should be officially declared a new geological age.

Despite these signs of gravity, there are some who believe that the Anthropocene is a passing fashion, a short-lived obsession, a bit of eye-catching jargon, and perhaps a political tool. It is argued that the use of the term sends a strong message to the public about the scale of the impact of Man on the environment[13]. Perhaps the greatest impact is on atmospheric concentrations of carbon dioxide released by the burning of fossil fuels, from pre-industrialization levels of 270-275 parts per million, to about 380 ppm today (2013). If we use CO_2 concentrations in the atmosphere as a proxy for climate change, then we are currently experiencing a human-induced tipping point that deserves concerted action. Yet, despite the very real concerns about managing a changing climate, should we be in the business of establishing geological epochs on the basis of political motivations, however well intentioned?

The signs are that the Anthropocene will continue to be used as a powerful metaphor for the impact of Man on this small blue planet. To ground the term in a scientific appraisal will add force to its use. One hopes that the process of assessment does not take as long, and does not involve as much acrimony, as the great debates of the past over the subdivision of geological time and of the archive of rocks formed over those aeons. These debates have sometimes been controversial and bitter.

In the early nineteenth century the key players in the subdivision of Earth's geological record, and therefore of time, were members of the Geological Society, which was founded in 1807 in a pub in Covent Garden, London. In fact, Adam Sedgwick, Roderick Murchison and Henry de la Beche were all presidents of the Society at some stage. They differed greatly, however, in background. Sedgwick (1785-1873) was a cleric and professor at Cambridge, Murchison (1792-1871) was a wealthy gentleman of leisure, and De la Beche (1796-1855) was dependent on funding from the government through the Geological Survey.

In the 1830s, the counties of England were in the process of being geologically mapped. Murchison and Sedgwick had mapped Wales and had established the Cambrian (from Cambria, the ancient name for Wales) and the Silurian (from the Celtic tribe of south Wales, the Silures) systems. De la Beche was in the process of mapping the rocks of the county of Devon, some 100 kilometres to the south. A furious dispute broke out over the age of rocks in Devon, and their correlation with those in South Wales[14], which centred on the age of coal deposits found in Devon. The details need not concern us.

In 1834 Murchison attacked De la Beche's arguments, even though he had not visited Devon himself, which brought a sharp retort from De la Beche that Murchison should allow 'facts' to take precedence over his 'preconceived opinions'. Murchison and Sedgwick went to Devon in 1836, and discovered that De la Beche had made a mistake in his mapping, which meant that the coal-bearing rocks of Devon could now be correlated with the well known and much younger Carboniferous rocks of South Wales. De la Beche insisted that there was no significant gap in the rock record below the coal deposits of Devon, yet the older rocks looked so unlike their equivalents in South Wales, which are the deposits of ancient rivers called 'Old Red Sandstone'. What's more, the rocks in Devon below the coal deposits were found to contain fossils that made them distinct from Murchison's newly recognized Silurian system. In 1839, Murchison hit on the solution – the rocks in Devon were deposited at the same time as the Old Red Sandstone and the oldest rocks of the Carboniferous in South Wales, and should be recognized as a separate system, which was named, sensibly enough, the Devonian. The historian of science Martin Rudwick calls the whole affair 'The Great Devonian Controversy'. For Murchison this was the only logical conclusion but a painful one, since he had to abandon his previous view that the fossils in rock formations of the same age should be identical. Clearly, fossils varied from place to place, depending on the types of environments in which the organisms lived. This may seem obvious today, but in 1839 it was not.

The debate between these three great men of the embryonic

Geological Society may have been fractious, but was productive. Theories and pre-judgments certainly played a role, but eventually anomalous observations needed to be explained that could not be accounted for. The turning over of the original hypothesis allowed new insights that spurred on the investigation of previously unrecognized problems, and thus the great scientific machine rolled forward creakily yet steadfastly to its unknown future.

Much could perhaps be said about what was different in the blend of cleric, professional and amateur gentleman that made up the community of scientists in the early part of nineteenth century Britain compared to the juggernaut of the science industry of the present day. But we have digressed enough from our focus on the Age of Man.

By the time I emerged from dinner in a subterranean vault with other conference delegates, darkness had fallen softly and comfortably on the capital city. It glowed electrically as people scurried to their destinations, and traffic buzzed, screeched and hooted, leaving long neon streaks in the photographic plate of my memory. My taxi sped past the great park now lost in darkness as I returned to the quiet Kensington Square. Above the chequer-board of lighted apartment windows was an aircraft descending to Heathrow, identified by a blinking red light and a low sonic rumble, suspended in space as if held in place by invisible forces. I had time to reflect on the day's conference and the conversations I had had over dinner in the subterranean vault. Had it been a consecration of a Paradise Lost? Had we administered the last sacraments of a disappearing world?

> Earth felt the wound, and Nature from her seat
> Sighing through all her works, gave signs of woe
> That all was lost.
> John Milton, *Paradise Lost*, 1667

I reflected on the closing thoughts of one of the speakers, Erle Ellis from the University of Maryland, who said:

> 'In the Anthropocene, we must embrace humans not as destroyers of nature but as creators, engineers and permanent stewards of the biosphere'.

So the answer is 'No, it had not been a mournful consecration', because when Paradise is truly lost, it will not be the 'Age of Man', but the 'End of Man'. The blame game will then seem particularly futile.

The Stones of Venice

Travelling by train over and through the Alps, changing at the bustling Milano, and enduring a hot, crowded inter-city train across the Po plain, it was with a mixture of relief and excitement that we left Mestre and crossed the Venetian lagoon by colonnaded bridge.

The sediments delivered to the coast by the mighty Po River and its smaller neighbours have over millennia built a triangular protruberance nudging into the Adriatic Sea, leaving embayments north and south that have been transformed into lagoons by the formation of thin coastal barriers. Venice lies in the northern and larger of these embayments, surrounded by water, safe from invasion, a republic of tiny islands built of mud, silt and sand. The large lagoon is protected from the storms of the Adriatic by this coastal membrane of sand. But inside the windswept and wave-battered shoreline, the lagoon simply rises and falls sedately with the tide. The tidal range is less than a metre, but is large enough to perform the important task of flushing the canals of stale water with a new supply from the open lagoon. Choppy waves gently throw the tied-up gondolas about with a rhythmic motion that produces a clicking and flapping noise sounding like a heartbeat of this maritime city. Had the Venetians not driven piles into the mud and silt, the islands would have shifted over the thousand years or more that its sons and daughters have acted as a hub for nobility, art and power.

Today, the sedimentary conveyor belt that has sustained and advanced the position of the coastline is faltering, and the region is sinking under its own weight. Nature watches as Venice disappears sadly into her dotage, a dim reflection of her former glory, her marbled floors and priceless treasures left as a relic of her former beauty. As the art and architecture critic John Ruskin puts it in *The Stones of Venice*, she is

'A ghost upon the sands of the sea, so weak – so quiet, so bereft of all but her loveliness that we might well doubt, as we watched her faint reflection in the mirage of the lagoon, which was the City and which the shadow.' [15]

The train ride across the colonnaded bridge brought us abruptly to our destination on the edge of the Grand Canal. In the past, 'distance could not be vanquished without toil', but we did not on this occasion have the benefit of creeping up to the city by boat, allowing the emerging panoramas to work their magic on the senses. A few moments later we were sitting in one of a veritable armada of *vaporetti*, conscious of the smell of diesel, straining necks to gain a new vista, spectators in the real-life film set that is modern Venice.

Entering the perpetual twilight of the churches and palaces, we see in the vaults, walls and floors the proceeds of trade and foreign ventures, even the plunder of war. The ornamental patterns of the *pavimenti* seem to move like the waters of the lagoon. The oldest make up the floors of squares and the courtyards of palaces. They are made of a red brick, commonly in a herringbone pattern, framed with white Istrian stone – a limestone widely used in Venetian architecture, and quarried from close to the city of Trieste on the northern shores of Istria, valued particularly because it decays slowly, almost as slowly as does a marble. The bricks were manufactured in Venice, and especially at the mainland town of Mestre, the landfall of the present-day railway. These red bricks are porous, absorb salts and crumble if inundated by the waters of the lagoon, so it is not surprising that eventually they were removed from the biggest square in Venice. Bordered by the Basilica di San Marco and the Doge's Palace, and flanked by the column surmounted by a winged lion representing St. Mark the apostle, the red bricks were replaced by a grey volcanic rock called trachyte in 1723. This hard-wearing, grey-coloured trachyte, a silica-rich igneous rock, was quarried from the Euganean hills along the Sottomarina and Pellestrina coasts of the Adriatic Sea, just to the south of the Venetian lagoon, a step closer to the mouth of the Po river. Quarried trachyte therefore had a small distance to travel on barges to the city of Venice. Other trachytes were used, as in the square in front of the church of San Giorgio Maggiore, which stands monumentally on the island opposite St Mark's square, as if being cautiously eyed by the winged lion. This square platform, with its octagonal motifs of trachyte from Monselice (near Padua in the Veneto) and Istrian stone, directly abuts the blue waters of the Canale di San Marco. As in St. Mark's square, these slabs of trachyte and Istrian stone must endure the regular floods of the lagoon when wind and tide conspire with the sinking substrate and displace the pigeons from their stamping grounds.

Many of the old churches and palaces of Venice have decorative floors made of thousands of small pieces of coloured stone, arranged in artistic but geometrical shapes or depicting flowers and animals. Floors in Venetian mosaic contain a wild assortment of exotic stones, culminating in the minute detail of the *opus tessellatum*. The variety of textures and colours is stunning and the provenance of the stones is an illustration of the reach of the Venetians throughout the Mediterranean.

A range of nodular, red and pink marbles come from relatively close-by in Verona, close to Lake Garda along the northern limit of the Po valley. The Torri yellow marble comes from the same region, which came under the republic of Venezia in 1405. The almost flawless Carrara marble comes from the Tyrrhenian coast of northwestern Tuscany, quarried at first by the Romans after their conquest of Liguria. Black marbles come from Iseo in Lombardy – the mainland power much feared by the Venetians. A diffusely mixed cream and yellow brocatello marble comes from Siena. The white Trani marble comes from the Adriatic coast of southern Italy just south of the Gargano promontory that sticks out like the vestigial toe of a horse into the ocean.

Further afield, the red and pink nodular and brecciated Cattaro marble comes from quarries along the Dalmatian coast, a short journey by sea across the Adriatic. The pure white Parian marble, which rivaled Carrara for flawlessness, came from the Greek island of Paros in the Aegean Sea. The cipollino marbles, with characteristic green and white wavy streaks due to the presence of epidote and chlorite, likewise came from Greece. And finally, centrepieces in the *pavimenti* were often made of the deep blue lapis lazuli (from Afghanistan), with its shimmering specks of iron pyrite, the rich green of the copper carbonate malachite, and dark green masses of serpentinite. Repeated patterns were made of black and white marbles, porphyry with hues of red from ruby to cherry, garnet and purple, and a Mori yellow-pigmented powdered limestone.

You have by now gained an impression of the wealth of colours and textures of the fabrics that make up the stones of Venice and their very wide provenance. You can, I hope, imagine the floral designs with petals of flaming yellow Verona marble, and Alpine green marble-breccias as the stems and leaves. You can imagine the strict geometrical forms of the grey masegno trachyte blocks framed with white Istrian stone in open quadrangles beside the water's edge. And you can imagine the floors of stars, diamonds, circles and squares with dramatic centre-piece roundels of ruby-red porphyry, green Alpine marble, precious malachite or lapis lazuli, surrounded by alternating wedges of black, white and salmon-pink marble. It is all a sensory overload.

Venice stands at the historical meeting point of architectural styles, cultural traditions and religious philosophies. She also stands as a celebration of the use of geological materials, like a living museum of Earth history, which helps to preserve her loveliness. But for many visitors, she is the city in the lagoon. For you can see plenty of red corrugated roofs and cathedral facades in metamorphic green and sugary white limestone in Siena and Florence, but you cannot sit on trachyte steps and stare across the bright waters of St Mark's canal at a winged lion anywhere but Venice. Her music is the sound of the lagoon against her curvilinear edges, the lapping of wavelets and the rocking of tied-up gondolas. It is the slow rhythmic beat of long oars entering the green, shaded waters of small canals between vertical cliffs of crumbling plaster.

Perhaps this is what signifies the Age of Man. It is not doomsday or Armageddon. It is the quiet and tenuous merger of Man and Nature. Venice may be in her dotage, but she is also a symbol of this image of the Anthropocene.

In 1991, before the current interest in the Anthropocene, British Earth scientist Euan Nisbet wrote in *Leaving Eden*:

'The path to a new global economy will be long and difficult, but on the horizon for the first time it is possible to see the faint image of humanity at peace with itself, and at peace with its environment.'

The future will tell if Nisbet's faint image of humanity becomes more decipherable.

6 THE GREAT SNOWBALL FIGHT

*In which contrasting and passionately held scientific opinions are based on
the weighting given to the same evidence . . . .*

The Jewel of Arabia

Years of working in the Alps and Pyrenees of western Europe did not
prepare me for the experiences of fieldwork in the mountains and deserts
of the Sultanate of Oman, the jewel of Arabia. We were there at the
invitation of the national oil company, Petroleum Development Oman,
who explore for oil and gas in rocks that are found scattered amongst
desert dunes and arid mountains, but which plunge beneath the surface
where they hold their carbon treasure. Despite an active and forbiddingly
expensive exploration programme, little was known about the deepest and
oldest of Oman's sedimentary rocks, apart from the information gained
from field expeditions in the 1960's conducted by men with double barreled
names and good Oxford educations, and from exploration by drilling by a
younger generation of pioneering Shell geologists. I therefore turned up in
Muscat with a colleague, a graduate student and a field assistant, ready to
find out more.

Coming from the greyness of an Oxford January, with that deflated
feeling following Christmas, I was struck by the sheer brightness of the
place. Whitewashed walls are splashed with the vivid purples and pinks of
bougainvillea; date palms, agave and yucca are everywhere; and the Indian

Ocean is the deepest of blue against the white limestone cliffs flanking the small bay where we stayed during our first few days. Touristic jaunts revealed that in old Muscat, where the modern sprawling city originated as a small fishing port, the white limestones are replaced by dark brown splintery rocks that have been heaved to the surface of the Earth from depths of tens of kilometres. They form an imposing and exotic backdrop for the port and in no small way contribute to a feeling of its strangeness – the same strangeness that an Omani would no doubt feel if he were walking through some cobble-stoned alley in Oxford on a foggy November morning, kicking through the beech leaves, fallen like amber, and dodging the impatient cyclists.

Other first impressions were daunting, as we contemplated why we were there. We were treated to stories of vehicles stuck in a glutinous, wet silt known as *sabkha*. We were warned of the perils of loose sand dunes that bury campsites overnight. We were in partial disbelief at the stories of the camel spiders that lurk in the dark eager to pounce on the slightest area of unprotected human flesh. And we were required to take advanced courses in off-road driving given by an ex-military sadist.

We explored long wadis incised deeply into towering and unstable mountains, with gravel floors lined with dark green date palms watered by miniature concrete canals that run at a carefully controlled gradient from the springline to the irrigated terraces. Immense boulders, and a lot of other debris, are periodically moved along the floor of these wadis during torrential floods, but our concern was to drive round these obstacles in search of ways of access to the dark cliffs in the hazy distance. These cliffs hold the mystery of time in their ravines and slopes; the rocks making up the cliffs were formed as mud, sand and pebbles on an ancient sea bed, so ancient that the Earth had barely awoken from her microbial slumber. We learnt fast, stumbling at our own ignorance and sometimes being blown off course by the previous stumblings of others, including the well-educated Oxfordians. We were in the Jabal Akhdar - a simple fold of rock, more than a kilometre high, its centre hollowed out as if by an ice cream scoop to reveal the object of our interest, surrounded by the chocolate brown rocks of inner Earth.

In the distance the dry, featureless plain extends to the coast, where wells extract water that fell as rain on the Jabal Akhdar ten thousand years ago - time capsules of raindrops, falling, gathering, flowing and seeping on a ten thousand year journey, to be extracted and consumed in one day by the thirsty and growing population of this eastern corner of Arabia. They do not know where the water will come from to quench the thirst of populations ten thousand years in the future. I think they care about that, but perhaps not enough.

In later years, Petroleum Development Oman funded our work through the 'Five Professors Project'[1] - not dissimilar to the three musketeers in some ways. Memories of our exploits, particularly in the field, are as fresh, as vivid, as if they took place yesterday, just as the story told by the ancient rocks we were investigating seems to be an item of current affairs rather than a reconstruction of mother Earth in deep time.

We found evidence of ancient ice ages in the dark cliffs of the wadis. Year after year we returned, to those same mountain wadis, to the quiet, remote desert of the Huqf, and finally to the great escarpment of Dhofar, where summer monsoonal mists blown in from the Indian Ocean produce a strange annual greening. Each year our notebooks became filled with measurements and the codified hieroglyphics of our trade, and each year others got to hear about our discoveries, and thus the age of innocence was ended, and the great Snowball fight began.

Cold Blasts in the Past

If you view the Earth from an orbiting satellite, as we are now so used to doing that we think nothing of it, you see the dark blue oceans and green or light brown continents, both covered with swirls and speckles of white cloud. In the polar regions we are quite used to seeing the white expanses of ice, calving at its margins to send flotillas of icebergs and angular fragments of former ice shelf into the ocean. As familiar as this scene is, it is also mightily peculiar. The reason is to do with the stability of that much under-rated compound, the humble H_2O.

Among our solar system neighbours, only planet Earth has large co-existing reservoirs of water present as liquid, solid and gas. Compared to the total mass of the Earth, the percentage of water is not great, but it is present in all three phases. The oceans are covered locally by ice and consist mostly, obviously, of liquid water. The land surface is also locally covered with ice and contains liquid and frozen water in soils and permanently frozen ground, liquid water in groundwater, lakes, rivers and inland seas and even water bound up in plant tissue. The atmosphere contains water as a gas as invisible vapour, but also as water droplets and ice particles suspended in clouds. Taken as a whole, 97% of Earth's water by weight is in liquid form, nearly 3% is present as ice and a minute 0.001% is in the form of a gas[2].

In contrast, on Mars there is no known liquid water today, though in the past (in the appropriately named Noachian period) there were floods of water capable of cutting deep canyons that look eerily like those on Earth. The water probably erupted from the land surface by melting rather than falling as rain. The atmosphere of Mars is known to contain water vapour, and ice occurs in polar ice caps and probably as permanently frozen ground. Together, the north and south polar ice caps of Mars are slightly bigger in mass or volume than Earth's Greenland ice cap. On Mars, H_2O therefore exists near its triple junction[3], but liquid water is only possible at the lowest elevations, and elsewhere water exists near the boundary between sublimation of ice to vapour and deposition of ice from vapour[4]. On Venus, water vapour is found (at 20 parts per million) in the carbon dioxide-rich, suffocating atmosphere, but there is no liquid water or ice. The Venusian oceans have been boiled off and water exists far from its triple junction.

Having liquid water is probably critical for the habitability of planets for life. This is why the current (2013-2014) Martian exploration programme[5], aimed at the assessment of the habitability of that planet for life, is focused strongly on the evidence for liquid water in the Martian past. There is no reason to believe that there are not a very large number of planets in the universe that satisfy this condition of the presence of liquid water. The present state of Mars and Venus, however, serve as a reminder of what happens if we leave this comfort zone of three co-existing phases of water. Relatively small changes in temperature and water vapour pressure on Earth can cause the proportions of the different phases of H_2O to vary wildly. Those wild excursions include the possibility of ice ages. Water in its different phases is key to understanding climate.

The phases of water are dependent, in part, on global temperatures, but what sets the limits on temperature? It will not surprise the reader to learn that 99.98% of the heat received by the surface of the Earth is from solar radiation. Our convecting interior does its job of stirring things up beneath our feet, but is ineffective in transporting heat to the surface. In fact, the flow of heat from the Earth's interior is about the equivalent of a 100 Watt light bulb over an area the size of a tennis court. In contrast, we can instantly feel the effect of the Sun going behind a cloud, or wake up in the morning to discover a frost. Put simply, climate is driven by solar energy via our planet's external envelope of water.

We can carry out a simple energy balance for the Earth by working out how much energy we receive from the Sun and how much is lost into outer space. The energy received from the Sun depends on its temperature, and radiating as it does at a surface temperature of 6000 K[6], we can calculate both the amount of energy emitted by the Sun and the wavelength of that energy. When you observe sunlight it is bright, which is because the wavelength of the radiation is mostly within the range for visible light. The radiation received by the Earth depends on how far away we are from the Sun and the cross-sectional area of the Earth bathed in the Sun's rays. Some of this energy is reflected back into outer space from the Earth's atmosphere and ocean and land surfaces. The proportion reflected back is called the *albedo*. Light surfaces such as those of ice caps reflect a lot of radiation back, whereas forested land surfaces reflect back much less. The planetary albedo is on average 0.3, that is, the planet uses 70% of the incoming energy from the Sun in keeping its surface warm.

Carrying out the energy balance, we find that the average surface temperature of the Earth should be a decidedly chilly 265 K (-8 °C). Since we know that the average surface temperature is 288 K (15 °C), there must be something wrong or something missing in our calculations. What we have neglected is the blanketing effect of the Earth's atmosphere. Since the Earth is relatively cool, the wavelength of the radiation emitted back into space is long – in the infrared range. The gases in the atmosphere like carbon dioxide, methane and water vapour are effective at absorbing long wavelength radiation, but let short wavelength radiation (visible light) through[7]. This is responsible for the much talked about atmospheric greenhouse effect. The opacity of Earth's atmosphere to outgoing long wavelength radiation is responsible for raising the average surface temperature from a chilly 265 K (-8°C) to a rather pleasant 288 K (15°C).

It should now be clear that there are two main ways to change Earth's climate. One is to change the amount of radiation received from the Sun, through what are called astronomical factors. The other is to change the opacity of Earth's atmosphere, through what are called greenhouse factors. Earth's albedo will also change with the growth and decay of ice sheets, so acts in a feedback. The colder it gets, the greater the area of ice, which increases the albedo, which makes it colder still.

Astronomical factors are known to change the radiation received by the Earth from the Sun on time scales of tens and hundreds of thousands of years. These time scales are within the so-called Milankovitch band, named after the Serbian physicist who explained the different mechanisms responsible for variations in solar radiation received by the Earth. Changes in the opacity of the atmosphere depend on the fluxes of carbon dioxide, methane and water vapour, and are affected by the presence of clouds and aerosols. These changes can happen quickly. This is where it gets complicated, because there are natural processes causing variations in the greenhouse effect, which are superimposed by the impacts of humans. Carbon dioxide, for example, is released naturally by volcanoes, but also by the burning of fossil fuels. By looking back into deep time, we have the advantage that we can make the glorious assumption that human impacts are zero.

Is a glaciated world with polar icecaps a normal situation or is the present-day anomalous? The answer comes from the geological record.

Over the last 543 million years of the Phanerozoic[8], the Earth has experienced at least three major ice ages. The records of these ice ages are now in climate belts that are far from glacial, which at first sight is puzzling. However, the Earth's plates are in constant relative motion, so the present-day latitude of a continent tells us little about its joy-ride across lines of latitude in its past. For example, the record of a major glaciation that took place in the Ordovician-Silurian periods (450-420 million years ago) is now found in the deserts of North Africa and in the Andes of South America. A more extensive glaciation took place in pulses of ice advance and retreat during a long period in the Permian-Carboniferous periods (320-240 million years ago). Its glacial deposits are found as far afield as Antarctica, India, Australia, South Africa, South America and North America, having dispersed from a supercontinent located close to the South Pole at the time. Finally, there is the more familiar ice age in which we are currently situated. Ice built up in Antarctica about 36 million years ago, and the northern hemisphere became strongly glaciated at about 4 million years ago. There was most likely a less severe glaciation in the Jurassic-Early Cretaceous (170-110 million years ago). The Earth is clearly prone to a glacial climatic state, but overall, the Earth has slumbered in warmth for far longer than it has shivered. Over the time scale of the Phanerozoic, the Earth has become glaciated about every 150 million years.

The significance of this long repeat time isn't understood, and we will return to this troubling topic shortly. However, by focusing on the ice age of the last couple of millions of years, we discover that it was not a period of uniform coldness, but of fluctuating long cold periods (glacials) lasting about 100,000 years and shorter warm periods (interglacials) lasting on average 10,000 years. We are currently in an interglacial and, one might think, due for another cold blast. Conventional wisdom is that these geologically rapid fluctuations in climate are due to astronomical factors - the short-term variations in the solar radiation received by the Earth in the Milankovitch band.

The pioneering studies of recent glaciations were carried out on the European mainland, particularly in the Alpine region. Louis Agassiz (1807-1873) was a great Swiss polymath of the 19th century. His academic pedigree is interesting, since on moving to Paris for education, he came under the tutelage of Alexander von Humboldt, who taught him geology and botanical geography, and Georges Cuvier, who taught him zoology and palaeontology. Humboldt's belief in the 'unity of nature', by which is meant the holistic combination of biology, meteorology and geology, and Cuvier's work on catastrophic rather than gradual change[9] as the cause of the extinction of species, must have influenced the young Agassiz greatly. Although he devoted himself to palaeontology, and became an expert on fossil fish, he is perhaps better known for proposing that the Earth had once been subject to a great ice age. He got the idea after having brought to his attention the large blocks of rock that were derived from the Alps strewn around the slopes, and even on the summits, of the Jura Mountains, some 50 kilometres to the north. These erratic blocks could only have been transported by ice. He embarked on a study of glaciers, concluding that an ice cap had once extended over Switzerland. On a trip to Scotland with William Buckland, the professor of geology at Oxford University, he realized that similar deposits of glaciers extended over northwest Europe. Buckland had grown to seriously doubt his earlier idea of a great 'biblical' deluge to explain the same bouldery deposits. He now had a satisfactory explanation for them as the debris carried by former ice streams.

Although Agassiz set the foundations of the understanding of how glaciers moved and of the former extent of ice caps, it was Albrecht Penck and Eduard Brückner, in their 3-volume work called *Die Alpen im Eiszeitsalter* published in 1909, who established the idea of four great glacial advances north of the Bavarian Alps of southern Germany, separated by periods (interglacials) of mild conditions during which glaciers retreated. These stages were named after sets of river terraces - the Würm, Riss, Mindel and Gunz – which recorded the to-and-fro motion of ice cap advance and retreat. The early views on our most recent ice age were therefore derived from the extent of glacial deposits on land, in the plains stretching out below high mountain ranges, and involved a small number of advances separated by retreats. Four-fold phases of glacial advance were subsequently recognized in northwest Europe and North America.

What followed was a revolution, based not on data derived from land, but from under the deep sea. Ocean sediments, recovered in the Deep Sea Drilling Project, which became the Ocean Drilling Programme, commonly composed of tiny shells of animals that had lived in the overlying water, showed many cycles of fluctuating oxygen isotope composition (specifically the ratio of the heavier isotope ^{18}O to the lighter, isotope ^{16}O). This ratio is controlled by the combined effects of temperature and the extent of extraction of ocean water in ice caps, so is a sensitive indicator of climate change. The fluctuating cycles of oxygen isotope composition could be correlated closely with calculated levels of solar radiation received by the Earth[10]. These astronomical variations in energy were acting like a pacemaker for climate change on Earth. The marine record was then compared with the history of ice caps such as Antarctica and Greenland.

There is now conclusive evidence of the close relationships over the last million years between variations of solar radiation, temperature changes driving ice advance and retreat, carbon-dioxide concentrations in the atmosphere and sea level change[11]. A vast amount of effort has been spent in closely calibrating various indices, or proxies, of this high-resolution climate change. The result is a scientific near-consensus on the causes and effects of glaciation at the fine time scale of the last million years. However, this leaves open the controversial question of what causes the major ice ages with a recurrence interval of ~150 million years to initiate and end. Or expressed slightly differently, what causes the long-term alternation of 'greenhouse' and 'icehouse' conditions?

One explanation, though highly controversial, is that the solar system travels through the spiral arms of the Milky Way galaxy[12], which are like shards thrown out from a rotating Catherine wheel. During this journey the Earth is subjected to varying intensity of cosmic rays, which, it is claimed, affects global climate. The frequency of the passage through the spiral arms is quoted as 143±10 million years - strikingly similar to the repeat time of icehouse conditions on Earth. If climate on Earth is affected by the cosmic ray flux from the galaxy, then variation in the rate of star formation should be an important factor. Two such periods of enhanced star formation are suggested at 2000 and 1000 million years ago (2 and 1 billion), which might have been responsible for the ancient glaciations that are roughly of this age. If this celestial mechanism were responsible for major icehouse periods, it would introduce a new player in the climate game, namely the cosmos. There is therefore much to play for. Is this the beginning of a serious questioning of the conventional view of climate change, which elevates solar radiation and greenhouse effects to unassailable positions?[13]

Jan Veizer (1941-) of the University of Ottawa has collected a large number of measurements of oxygen isotopes from the carbonate shells of animals that once lived near the surface of the tropical oceans over the duration of the Phanerozoic. With co-workers, he compared trends in sea surface temperature derived from these carbonate shells with estimated variations in the cosmic ray flux on the one hand, and atmospheric carbon dioxide concentration on the other. He suggested that the link of climate change with CO_2 concentrations is sometimes weaker than the link with the cosmic ray flux[14]. The argument isn't simple, and goes something like this. The cosmic ray levels received by the Earth are controlled by the flux from the galaxy generated by exploding stars, but are affected by the shielding effect of the Sun's and Earth's magnetic fields. These magnetic fields act like an umbrella deflecting the cosmic rays to the side. Consequently, increases in the strength of the solar shield reduce the cosmic flux received by Earth.

Cosmic rays are thought to generate nuclei for the condensation of low-level clouds, which thereby induce cooling by reflecting more radiation back into space. In summary, variations in the flux of radiation from the cosmos affect Earth's climate. The cosmic or celestial hypothesis for climate change has caused a major dispute to erupt. Interestingly, the sharpest reaction has come from Germany, where Veizer also held a personal chair in the University of Bochum. The Potsdam Institute for Climate Change Research (PIK) took the extraordinary step of issuing a statement denouncing Veizer's work.

Arguments have been put forward to refute the celestial model for climate change for *recent* global warming. Opponents state that there is no reliable trend in the cosmic ray flux over recent time, and dispute the correlations (or lack of correlations) used by Veizer between global temperature and atmospheric CO_2 concentrations. If a celestial driver, rather than the greenhouse effect, increases temperatures, then what has caused the strongly increased CO_2 concentration in the oceans, atmosphere and probably biosphere since the 1960s? Although temperature regulates the transfers of carbon between atmosphere, ocean and biosphere, it is hard to believe that relatively modest global temperature rises alone could have driven the known strong increase in carbon dioxide concentration in the atmosphere. The hypothesis of a celestial mechanism involving variations in cosmic ray flux is therefore probably headed for the rocks when applied to recent climate change.

Just when it was thought that the idea of a link between solar activity and climate *via* the cosmic ray flux was dead and buried, the topic sprang to life in 2007 with the publication of a paper by Vincent Courtillot[15] making a connection between the strength of Earth's magnetic field and temperature variations in the last millennium. Convention, of course, is that temperature has been forced by gas and aerosol emissions *via* the greenhouse effect and that solar and terrestrial variations in the magnetic field are too weak to influence our flux of cosmic rays and therefore to affect climate. It sparked bad-tempered debate once again, with accusations that the data trends shown in charts were dishonestly put together.

Veizer and colleagues published their work in the ultra-competitive journal *Nature* [16] in 2000, and popularized the idea of a celestial driver for climate change in non peer-reviewed magazines (*GSA Today* in 2003, published by the Geological Society of America, and in *Geoscience Canada* in 2005). This tendency to disseminate the model in non-peer reviewed publications, however respectable, has attracted criticism. Veizer and colleagues have felt the full force of the climate science community for stepping out of line. Getting involved in scientific disputes is bruising and is strictly not for sissies. It remains to be seen if anything will be salvaged from the celestial model for climate change on Earth.

Other less controversial explanations for the long-term initiation and ending of icehouse periods are related to the motion of Earth's tectonic plates, uplift of continents, volcanic activity, changes in atmospheric composition or oceanic circulation. Although frustrating for those looking for a single mechanism, any of these possibilities may have occurred at different times. A joining of continental plates around the pole took place in the Permian-Carboniferous period to form the supercontinent Pangaea, but could not have affected the most recent icehouse period when plates have been dispersed. Uplift of continental plateaus such as Tibet may have been sufficient to tip the Earth into glaciation in the last 10 million years by affecting atmospheric circulation and by increasing chemical weathering and the drawdown of atmospheric CO_2, but it is doubtful whether this mechanism can explain more ancient icehouse periods since they do not have a clear association with mountain building. Volcanic activity releases the greenhouse gas CO_2, which promotes warming, and aerosols such as sulphuric acid that reduce it. Aerosol effects are short-lived, so increases in volcanic activity are unlikely by themselves to push the Earth into an icehouse period. Finally, changes in ocean circulation patterns caused by the opening of gateways and the forming of barriers by plate tectonics is used as an explanation of cooling leading to the most recent icehouse period, but it is unlikely that this applied to the icehouse period affecting the supercontinent of Pangaea.

With such disparate ideas failing to explain the periodicity of icehouse periods through geological history, it is perhaps worth giving the celestial model further thought. However, this hypothesis now carries a lot of baggage. Adherents are labeled as climate sceptics – and this is meant as a term of abuse, not to indicate sensible contemplation – or worse, as denialists. We therefore have an interesting phenomenon. A challenging and imaginative idea is dismissed when applied to the extremely well documented relatively recent history of glaciation on Earth amid accusations of a lack of honesty. This leaves the hypothesis damaged. Yet essentially the same hypothesis may be of use in explaining the periodicity of icehouse and greenhouse periods in deep time. It is too early to say whether part of the hypothesis will be salvaged in this way. It is a reminder of the unpredictable course of scientific debate.

The Biggest Deep Freeze

The great Snowball fight is about the idea of a glaciation deep in Earth's past that is alleged to have been global in extent, with temperatures falling to about -50°C for tens of millions of years. The original idea was put forward by an Australian geologist who explored the Antarctic, named Douglas Mawson (1882-1958). Mawson found rocks with the fingerprint of past glacial activity in South Australia, and came to the conclusion, wrong as it happens, that if glacial deposits were found in the low latitudes of Australia, ice may have covered the whole world[17]. Wegener's hypothesis of continental drift was ignored, plate tectonics had not yet been formulated, and Mawson did not realize that continents could move across lines of latitude. The past latitude of continents can be evaluated from the magnetism entombed in rocks of known age. When this technique was applied to rocks of glacial origin exposed in Svalbard (formerly Spitzbergen) and Greenland in 1964[18], it was found that the continent must have been situated in low latitudes at the time of deposition of its distinctive glacial sedimentary rocks. The idea of a global glaciation was therefore given new life. Brian Harland of Cambridge University termed this the great Infracambrian[19] glaciation, but he was careful to suggest that the glaciation was 'widespread' rather than 'global'.

Brian Harland (1917-2003) was an extraordinary man and a fellow of Gonville and Caius College, Cambridge University, to which I was admitted as a doctoral student in 1976. For the first two years of my doctoral studies, I was given workspace in the Spitzbergen room, at the very top of the Geology Department, otherwise known as the Sedgwick Museum. During the winter, it did indeed feel like Spitzbergen. The Cambridge geology department in those days was in a state of suspended animation, though it didn't seem that peculiar to me at the time. The internal doors between the Sedgwick Museum and the Department of Mineralogy and Petrology were locked, and the geophysicists who were making groundbreaking advances in marine geophysics and plate tectonics were never seen, since they occupied rooms a couple of miles away in a country villa. Morning coffee and afternoon tea were held in the ground floor of the Museum, complete with marble inlaid tables and overlooked by busts of eminences set between glass cabinets containing polished slabs of building stone. It felt as if I were in a Precambrian academic world, but the expert on the Precambrian (or at least the most recent part of it, the Neoproterozoic) resided in a paneled room that directly opened into the mausoleum with the stone eminences.

I wish that I had understood what Brian Harland was doing in those years spent as a doctoral student, but I was too immature to make it my business, and Brian was too busy to worry about 'outreach'. Brian Harland was into painstaking scholarship and didn't give a hoot for outreach, or citations, or, judging from his appearance, for anything remotely to do with the world and its creature comforts. Perhaps this was something to do with the fact that he was a Quaker. His former young protégé Ian Fairchild helped to complete Brian Harland's last, unfinished paper after his death in 2003. I dissected this paper[20] for traces of a 'road to Damascus' conversion to the idea of global glaciation and could find none.

Shortly after Harland's documentation of former low-latitude glaciation in the Arctic, a Russian physical climatologist named Mikhail Budyko (1920-2001) conducted a back-of-the-envelope calculation of the energy balance of the Earth. He found that by balancing the in-coming radiation by the out-going radiation, and making use of the changing albedo of the Earth's surface caused by the growth of highly reflective ice, there would be a positive feedback loop if ice grew as far equatorwards as the tropics. After this tipping point, there would be runaway ice growth leading to a global freeze[21]. Now it is not that remarkable that a planet might refridgerate, as a casual consideration of Mars will confirm. And for that matter, it's not that remarkable for a planet to turn into a boiling inferno, as in the case of Venus. But what is odd is that once encased in ice from poles to equator, what could cause our fickle planet to go into reverse gear and start melting? For tens of millions of years, it is claimed, the Earth was lapt in the universal law of a frozen shutdown. There was no weather, no clouds, and certainly no barbecues, just the rising of the moon above the icy wastes but observed by not a single living creature. Budyko regarded the prediction of a global glaciation as a theoretical exercise, since according to his calculations, it would be impossible to escape from this global refridgeration. He concluded that global glaciation *could* happen, but didn't.

Budyko's calculations left out an important effect that might solve the problem of the escape from global glaciation, namely the release of carbon dioxide from the interior of the Earth through volcanoes and vents. Acting as a greenhouse gas, the build-up in carbon dioxide levels in the atmosphere above the frozen Earth would, according to the Caltech geobiologist Joseph Kirschvink[22], lead to a supergreenhouse, promoting catastrophic melting of the planet's ice sheets. Kirschvink called the state of global glaciation 'Snowball Earth', and with this epithet introduced into parlance, the scene was set for an explosion of activity in an attempt to test the validity of this seemingly outrageous but compelling idea. It has given the Earth sciences one of the most colourful dogfights in recent history. The dogfight shows the power of differing world-pictures, and the Snowball Earth metaphor is one of the most obvious dramas of our time.

The Snowball fight is about a scientific dispute, but it is far from the sterile turning of the wheels that some believe the relentless progress of science to be. Science does not proceed through the faultless connection of facts by the invisible strands of reason. Instead, the great Snowball fight is essentially about a a brilliant man with flowing grey hair, intense eyes and a restless mind, who brought passion back to the understanding of Earth's long narrative. He is an iconic man, with a quick, ascerbic tongue and slow, awkward apologies, and with a multitude of obsessions from jazz to baseball to marathon running. The popularization of the idea of Snowball Earth is due in large part to his tenacity and insight.

Paul Hoffman (1941-) recognized that glacial deposits in Namibia were sharply overlain by unusual carbonate rocks with carbon isotopic ratios[23] that suggested a cessation of biological productivity in the world's oceans. He concluded that this was powerful support for the Snowball Earth idea, and realizing the significance of this finding, published the first integrated account of the theory in the highly respected journal *Science* in 1998[24] and the widely read magazine *Scientific American* in 2000. The Snowball Earth idea immediately hit the media, with television documentaries, many column inches in newspapers and science magazines, website opinion pieces, and a popular science book entitled *Snowball Earth* by Gabrielle Walker published in 2004. It was not until 2002 that a fully considered hypothesis was published[25], after the initial media splash, and from that date to the present, the pace of activity for and against this startling idea has been breathless. This can be gauged by working out the times 'Snowball Earth' has been used in published journal articles using 'Web of Science'[26]. The number of occurrences increased from little more than zero in the late 1990s, rising steeply to reach a maximum of 79 in 2007, and then falling slightly to the present day.

Academic scientists are very aware of the uses and abuses of statistics generated in databases such as 'Web of Science', since academics are commonly assessed on the basis of the number of citations of their work by others. Getting a job in the first place, and being promoted if you already have one, is strongly influenced by the scientist's citation statistics. It is a current fad, widely used by bureaucrats and managers, and sadly by research grant committees, who need some metric for assessment and who would find it impossible to dive into the original published works and read them. Yet a citation to my published work that says that I am completely and utterly wrong scores the same as a citation that declares that I have made a valuable breakthrough. If you want a highly cited article therefore, it is best to write something that is contentious or provocative, since the worst thing of all is not to be disagreed with, but to be un-noticed. This gives science a tendency to concentrate on short-term gain by publishing attention grabbing, speculative papers. There is a difference between a first-class piece of work that can explained in simple terms to a barmaid[27], **and a piece of work that looks as if it has been written by Homer Simpson.**

Part of the reason for the relatively heavy citations of 'Snowball Earth' is because it is contentious. However, there are further social factors that are at work. Opponents of a new hypothesis commonly spend some time considering the scientific problem and publishing their findings, but eventually walk away, preferring to spend time on more fruitful avenues of research. There is therefore a wastage of scientists pulling away from front-line battle, replenished by new recruits with different agendas, and different ideas to test. The citations derived from 'Web of Science' show that Snowball Earth continues to be an engaging idea, whether it is agreed with or not.

For several years, we tested this idea of global glaciation by studying rocks deposited during putative Snowball Earth events in Oman, and eventually concluded, like party poopers, that in our opinion, the idea was flawed[28]. We were convinced that we had found clear evidence that during these ancient glaciations, the ice advanced and retreated in cycles of colder and warmer climate, much as is known to have happened in the last few millions of years.

Whereas the hypothesis invoked a global shutdown of the water cycle, we found strong evidence in favour of dynamic glaciers. Others suggested that the distinctive carbon isotopic values of carbonates deposited prior to and immediately after glaciation did not unambiguously point to a shut-down of productivity in the world ocean. Yet others suggested that the carbon-dioxide concentrations in the atmosphere at the time of deglaciation were little different to those of the present-day, rather than at 300 times present-day values required to produce a supergreenhouse.

The hypothesis remains controversial. Many numerical modelers of past climate make an assumption that ice cover was global, and thereby generate computer simulations that satisfy their intellects but which may be simulating something that never happened. Geochemists plough on regardless, doing the same. Field geologists commonly ignore numerical modelers, and numerical modelers scratch their heads wondering which geologists to believe. There is probably no silver bullet that will resolve the debate. We take different sides largely because of our world-pictures. These world-pictures of how things should be tug us in different directions. No one is trying to mislead anyone else. All convictions are held honestly. Different scientists weigh their own observations more heavily than those of others. What is highly persuasive for one scientist is ambiguous or bogus for another.

On the Mountain of Truth

It would be surprising if the conflict between inductive and deductive approaches did not rear its ugly head in the case of the Snowball Earth hypothesis. This tension was not consigned to the scrap heap of history after the mauling by American scientists of Alfred Wegener's hypothesis of continental drift. The grand theory is this time global glaciation, not the mobility of continents. However, the dynamic is different this time round. In Wegener's case, the basic observations were sound, but the grand theory seemed lofty and impossible to find a mechanism for. This is not true for the Snowball Earth debate. Instead, geological observations are in tension, but the theory is not dismissed by virtue of its loftiness or strangeness. The Snowball Earth idea is more a good example of abduction – the finding of the best explanation that works through a creative act of finding hitherto unconnected observations.

Geologists these days are unlikely to be surprised by anything except for a spontaneous outbreak of world peace. Many field observations, collected not aimlessly, but in a targeted fashion in order to directly test the hypothesis, are at odds with it. If we inhabited the world of physics, we would discard the hypothesis and start again, but instead we inhabit a house of underdetermined science. Consequently, the global Snowball Earth idea that we started with has morphed into a softer hypothesis that allows for small bodies of ocean water, for ice movement, and for refugia to allow life to survive these dramatic challenges of extreme climate instead of 'death enveloping all Nature in a shroud'[29]. This is satisfying for some, since it enables them to inch their way laterally. But as the morphing process continues, the question arises, how different were these great glaciations of the past from those of the Ordovician (420-450 Ma), the Carboniferous-Permian (260-360 Ma) and the Pleistocene (2.58 Ma to present) periods?[30]

And if the glaciations weren't global in the strict sense, then surely the mechanisms proposed by the originators of the idea must fall by the wayside. In short, what of the original idea is left?

Our hypotheses in science are not like our young children, about whose intelligence and good looks we can be proudly biased. Science progresses by disputation, by argument against argument. The problem with stories about the history of the Earth in deep time is that they are difficult to disprove or prove to the satisfaction of all parties. This is what happened with the great Snowball fight. All sides in the debate find it difficult to agree on a binding test of the idea. There are arguments pro and con, and each side from its own vantage point claims ascendancy for its own evidence.

In 2005 I organized an international conference on the Snowball Earth idea on the aptly named Monte Verità, the mountain of truth, which stands above the lakeside towns of Locarno and Ascona in southern Switzerland. The Lago Maggiore extends into Switzerland from Italy, its bed cut deeply into rock by former glaciers, the depression now filled with blue, shimmering water. In former days the Monte Verità served a range of cultural needs, most of them Utopian. The hill was initially purchased in 1900 to be used as a co-operative vegetarian colony, initially run on socialist lines, and later became an establishment promoting strict nudism and the rejection of property ownership, marriage and party politics as well as the wearing of clothes. It became a centre for the expression of arts, women's rights and mysticism.

The present conference centre is owned and administered by the federal university of Switzerland (ETH) and was available to me as a professor at the Zürich campus. It was built in 1927 in the modernist, 'total' architectural style of the Bauhaus. Several years later the Bauhaus school was closed by the Nazi regime, accused of being the centre of intellectual communism, but its style in architecture, art and design nevertheless spread throughout the world. For 5 days we debated in the wonderful Bauhaus conference centre on the mountain of truth. Each evening I retired to my room, stood on the balcony overlooking the wooded slopes descending to the blackness of the lake, and held to the hope that there would be a meeting of minds, or a clear indication of the way forward, but there was no consensus forged and no agenda agreed. I do not know of one person who changed their mind at the conference.

I no longer go to the jewel of Arabia, but I occasionally reminisce on the Five Professors Project and on the eerie peace of our camps in the desolate wilderness of the Huqf. There was a low hum that broke the desert silence in that place; it seemed to be drawn from somewhere deeper, as if originating in the very rocks and dunes that surrounded us. How small I felt, sitting on a rocky platform at sunset, with the wind gently rippling the sleeping bag as a sea breeze flowed from the Indian Ocean across our faces.

For me, the great Snowball Fight is over. For a while I intersected with a great man with flowing grey hair. Passions have subsided and adrenalin has been put to alternative uses. Does it really matter whether planet Earth froze over completely and profoundly, and had the audacity to do it at least once more, deep in her microbial past, like an out-of-control, binge-drinking teenager? It frames one's view of how resilient the Earth is when pushed to the limit, and of its ability to recover to spawn the evolutionary processes that led to you and me. Our findings in the desert of Oman suggest that teenage Earth was not quite as rebellious as thought - we had simply discovered her mood swings. We are left with an abiding image of a resilient, complex, nurturing and forgiving Earth, not the reckless and temperamental planet of alarmists' dreams. And that, perhaps, matters.

7 BLIND DATES: THE ANTIQUITY OF THE EARTH

In which natural science slowly converges on an answer over a period of 300 years, as new insights and technologies become available . . . .

In Marble Halls

What could be more fundamental to know than the age of the Earth? Before the flood of knowledge that marked the Renaissance there was little need to obtain information on the age of the Earth and the cosmos. When the Earth's antiquity began to be questioned, the Judaeo-Christian Bible was turned to for a solution to the problem since there were no other ways of knowing. Many scholars made such estimates[1], but the best-known and most exact attempt is that of Archbishop Ussher (1581-1656), Archbishop of Omagh, Primate of All Ireland and fellow of Trinity College Dublin.

Trinity College Dublin was founded by Elizabeth I in 1592 (not by Elizabeth II as I was informed by a helpful taxi driver on my first visit in 1996). Trinity is a College set in the heart of Dublin behind classical facades, high walls and wrought iron railings. It is entered through a gated bit of Renaissance architecture that is beautiful yet instantly forgettable, as if it were created formulaically by a robot with a slide rule instead of by a creative genius. But I am sounding like a latter-day reincarnation of John Ruskin[2], so will get to the point. The spacious front square, known as Parliament Square, is bordered by a variety of buildings; the Chapel that stares at the Public Theatre opposite, and along one side stands the majestic

Campanile on the edge of an ancient lawn. The Old Library, which contains the famous Book of Kells, was founded at the same time as the College by Bishop Ussher, who donated many priceless printed books and manuscripts. Beyond it lies the quirky Museum Building built in the Romanesque style and finished in 1875, which houses the departments of Geology, Geography and Civil Engineering.

From the outside the Museum Building makes you feel as if you are looking at a Venetian palazzo – all that is missing is the choppy waters of the Grand Canal and a few moored gondolas. As befits the home of a geological collection, the building is made from a large number of different stones, including examples from every quarry in Ireland operating at the time of its construction, supplemented by white Portland Stone from the Isle of Portland, England as a horizontal string course, and Caen Limestone from northern France in the semi-circular tympanum above the heavy front door. There are 108 columns on the exterior, each surmounted by a carved capital. The carved capitals are mostly in the form of flowers and speak of the beauty of the natural world. Inside, the cavernous, domed hallway leads to two symmetrical staircases made of Portland Stone. Irish marbles and a dark red serpentinite from the Lizard area of Cornwall make up the columns of the entrance hall and the sloping flat slabs serving as a coping above the stair balustrade. High-spirited students have been known to slide down these smooth marble slabs, and only an elevated level of decorum has prevented members of staff from doing the same.

The Museum Building must be the most impressive home for any Geology department in the world. The architects were the Irishmen Deane and Woodward, and it will perhaps come as a surprise to learn that these architects also designed the University Museum building in Oxford, where Huxley debated with Wilberforce in 1860. One building is Byzantine, the other Gothic, but the carved capitals give away the identity of their designer.

Bishop Ussher's portrait, however, isn't found in this Byzantine gem[3]. Instead it adorns the Examination Hall off Parliament Square.

When the sun slants across Trinity's dewy lawns,
and trees in the throes of resignation
before the onslaught of winter's purge

blaze with chiaroscuro light
and surrender gladly to ancient laws;
When the damp autumnal air
clings to grey Renaissance squares,
invades the Museum's byzantine hall,
and blurs the Campanile's triumphant sound,
I think of you.

When tangled heather takes on a deeper hue,
from pink to plum to purple-red,
in matted form of twisted roots,
and ink black peat pierced by quartzite slabs
veneers the landscape's fragile skin;
And when the hillsides' parabolas,
give way to noisy waterfalls,
in steep descent of silver rain
across ancient Ireland's faulted lines,
I think of you.

When above Bray's headland, dark and solemn,
hover clusters of orographic clouds,
as if attached by wires to Earth below,
and the pointed sugar loaves guard,
like sentries Dublin's southern flank;
When shafts of light emerge from
leaden skies over Wicklow peaks,
a rising breeze causes pace to quicken,
and bearings turn for homeward bound,
I think of you,
And smile[4].

Ussher published his work in Latin in two parts in 1650 and 1654 (it was translated into English in 1658), concluding that the Earth was created on 26 October 4004 BC. He worked this out by counting back all the generations from the present to the first humans, using references in the Bible. Ussher's estimate is best known because it was inserted into the marginal notes of some editions of the 17th century King James translation of the Bible. Poor Archbishop Ussher - he has been lampooned as some kind of lunatic in the light of much later estimates of the age of the Earth. It is easy to pour scorn on the tradition of biblical chronology, but these scholars and naturalists merely codified prevailing views of the natural world.

The prevailing views of the world were based on a reading of the book of Genesis in the Judaeo-Christian Bible. It is instructive to find out what this book actually says, and to ask whether it should compel its reader to believe in a very young Earth. Is the Bible a library book of history that can be used in this way? Was it 'written as a handbook to understand Nature', as Steve Jones states in *The Serpent's Promise*? I think in both cases the answer is 'no'.

The Bible states that in the beginning God created the heavens and the Earth out of nothing, or *ex nihilo*. Initially the Earth was formless, empty and dark, but God said 'Let there be light', and light was separated from darkness. The Old Testament of the Bible often uses 'God said' to mean that God acted. This act therefore creates day and night out of the primaeval darkness. The section concludes with a formula that recurs throughout the passage: 'And there was evening, and there was morning – the first day'. Note the order – evening, then morning. In Hebrew, the word for evening (erev) also implies disorder or chaos. The word for morning (boker) also implies order[5]. So the way in which this phrase reverberates through the passage is constantly reminding the reader that God created order out of chaos in each of the great acts of creation. The use of this recurring phrase may have been a device to help with the oral transmission of the Genesis story.

God separated water from water by creating an 'expanse' or atmosphere which he called 'sky'. God then separated the water from dry ground and called the dry ground 'land' and the gathered waters 'seas'. The land produced vegetation – seed-bearing plants and trees bearing fruit with seed in it. God then put lights in the sky – a greater light to govern the day, which is of course the Sun, and a lesser light to govern the night, which is the Moon. He also made the stars. It is relevant that the Sun and Moon are not explicitly mentioned, perhaps because they were the names of pagan deities. Both the greater and lesser lights were clearly placed under the greater power of God.

God then created the living creatures of the seas and the birds of the sky, and declared that they should flourish and be fruitful. 'Winged birds' refers to all flying creatures, including insects. Next, God caused the land to produce creatures – both livestock and wild animals. He then made

Man, who is effectively given stewardship for the world in which he lives.

The themes in the creation account, and the pattern of Genesis Chapter 1, are very similar to other Mesopotamian creation stories, such as the *Epic of Atrahasis* (written about 1600 BC) and the later *Enuma Elish*. There seems little doubt that the early chapters of Genesis have their roots in Mesopotamian sources, but with heavy revisions in the light of Israel's emerging distinctive theology. Many of these early mythical stories from Mesopotamia involve the theme of a great conflict between a deity and a sea monster representing watery chaos. In the Babylonian creation myth, the supreme God of Babylon Marduk defeats the female sea dragon Tiamat. Creation involving separation, such a striking feature of the Genesis account, is also found in early Sumerian texts and the Babylonian creation epic, where Marduk defeats Tiamat, cutting her in two, making one half of her corpse into the sky and the other half into the Earth.

Yet despite these borrowings, the Genesis account not so much *imitates* these other creation stories as deliberately *rejects* them. It provides a very different world-picture of one creator instead of many; it places the sun, moon, stars and sea creatures, not as gods but as part of a single God's creation; it demonstrates that the world is a planned and orderly, majestic creation, not some kind of cosmic accident; it places man at the apex of creation, rather than with a subservient walk-on part; and emphasizes that man has a special position in creation, made in God's likeness (self aware, possessing a conscience, moral, creative, spiritual), with special responsibilities towards the created order.

The purpose of the author of Genesis Chapter 1 is clearly not to explain creation, in the sense that a scientific theory attempts to explain the temperature-pressure-volume relationship of a gas for example. Instead the author wants to catch us in awe and wonder. By trying to squeeze explanations of the natural world out of the text, we may be trying to wring water out of a dry shirt.

The essential structure is of 6 days of creation, with the regular refrain 'and there was evening and there was morning' moving the story along, perhaps suggesting that the story was originally a hymn or a poem, sung or recited at festivals. There is a second level of structure. The first three days produce form out of formlessness, that is, they are acts of 'separation' – the separation of light from dark, the separation of water in the oceans and in the atmosphere, and the separation of land from sea. The second three days then fill what has been separated. So we firstly fill the heavens with the sun and the moon (Day 4), we fill the sea and the air with living creatures (Day 5) and we fill the Earth with fertile vegetation and finally Man (Day 6). That there is clearly this structure of three days of forming and three days of filling should immediately warn us not to take the 6 days in strict chronological order.

However, writers attempting to retell the Genesis account as science overlook this literary (not literalist) structure and in so doing get impaled on the thorn of chronology. The geneticist Steve Jones[6], for example, writes that the Sun's rays were necessary for turning the world green, which caused a build up of oxygen and eventually the evolution of higher animals, and adds (page 84, *The Serpent's Promise*)

> 'Genesis has a different view for the Lord commanded 'Let the earth bring forth grass' even before he created the Sun'.

In a similar vein, he writes (page 70)

> 'The biblical version has: 'And God said, let the waters bring forth abundantly the moving creatures that have life, and fowl that may fly above the earth in the open firmament of heaven', which brings the birds in too early'.

A critical requirement for examining the truthfulness of the Bible as a history book should be an understanding of the symbolism and literary devices used by its authors. Otherwise, scientists are trying to demolish an edifice that never existed or was never intended in the original writings. The Bible is not a history book, or even a library book. It is more a book for the pulpit.

There are several schools of thought on the Creation story[7]. Young Earth creationists believe in a literal six days each comprising 24 hours. At the other end of the spectrum, others believe that the account is entirely literary rather than literal. In this latter view, the schematization of the Genesis account is interpreted as a literary device used to stress the system and order built into creation, and is therefore not to be taken literally or chronologically.

Early Christian leaders in the Greek and Roman cultures already took the allegorical view that the 'days' of Genesis were 'ages' of unknown duration. Origen, who was perhaps the most important theologian and scholar of the early Greek Church, wrote in AD 231 ridiculing the idea that the 'days' of Genesis should be taken literally. Despite this, twentieth-century biblical literalists such as E.J. Young, H.M. Morris and E.H. Andrews are all insistent that Genesis Chapter 1 is chronological. We have the surprising fact that the wisdom of early writers has been superceded by a later, narrow fundamentalism - an example of what Dick Taverne would call the 'March of Unreason'[8] and Steve Jones would call the 'Age of Endarkenment'[9].

Doubts on the Biblical Chronology

Doubts about Archbishop Ussher's biblical chronology soon surfaced at the end of the seventeenth century[10]. Very little escaped the attention of English physicist and mathematician Isaac Newton (1642-1727), who carried out a thought experiment[11] in 1687 on the time required for a sphere of molten iron to cool. When scaled up to the dimensions of the Earth, he arrived at a figure of 50,000 years.

At the same time, new discoveries of fossils were transforming the mental landscape and pointing to an antiquity of the Earth that extended back before 'even the very pyramids', according to English natural philosopher Robert Hooke (1635-1703). Fossils were seen as the route to a chronology of the Earth. Their presence at high elevations on mountaintops convinced the French anthropologist Benoit de Maillet (1656-1738) that sea levels had been declining progressively over time by evaporation of ocean water[12]. This would, he estimated, take 2 billion years to reach today's levels – which must have appeared close to eternity for the late seventeenth century mind set. De Maillet had no idea of tectonics, which caused sedimentary rocks containing marine fossils to be uplifted in mountains, and he did not realize that sea levels went up and down periodically, for example in response to glaciation and ice melting.

By the late eighteenth century, the age of the Earth became a subject of sustained and public debate in both England and France[13]. Georges-Louis Leclerc, Comte de Buffon (1707-1788) thought that the planets of the solar system formed when a comet hit the Sun. Initially, the Earth must have been molten, he argued. He must surely have known of the calculations of Isaac Newton, because he conducted experiments in a foundry with balls of molten iron. He calculated that it would take nearly 100,000 years for the Earth to cool to its present state with solid rocks beneath the land surface and liquid water in the oceans. The significance of this calculation is that it is significantly longer than the biblical age suggested by Archbishop Ussher.

In Britain, it was dawning upon geologists that the time required to explain cycles of uplift and deformation by tectonics followed by erosion and deposition of sedimentary rocks, was very large. A contemporary of Buffon, the Scottish geologist James Hutton (1726-1797) referred to the vast expanse of geological time in his phrase 'we find no vestige of a beginning, no prospect of an end'[14]. Hutton's work did not allow an age of the Earth to be calculated, only to suggest that it was indefinitely large.

Interestingly, revolutionary ideas of a much older Earth than envisaged from the Biblical chronology seem to have arisen at a time when France was undergoing its own revolution of a political nature. The ideas of Buffon in his *Epoques de la Nature* (1778) in France and of the geologist James Hutton in his 2-volume *Theory of the Earth* (1795) in Britain were countered with a renewed thrust to assert the priority of the biblical chronology. An example is the Irishman Richard Kirwan (1733-1812), president of the Royal Irish Academy as well as Fellow of the Royal Society, who dismissed Hutton in his *Geological Essays* (1799):

'the function of geology graduates into religion, as this does into morality'.

Kirwan's attack on Hutton was both misguided and intemperate, but at least it stimulated the great Scottish geologist to respond, forcing him to hone his views. Hutton's ideas were to become the foundation of much of modern geological thought. This little cameo illustrates a point that is often neglected by the general public. Scientists are warm-blooded human beings who spar with each other, sometimes quite rudely. The business of sparring is part of the way in which ideas are refined. Scientific conferences are very dull affairs unless participants are prepared to stick their necks out and engage in a lively and serious debate about their findings. The progress of science therefore thrives on disagreement and controversy, out of which eventually filters generally accepted views on the way that Nature works.

At the same time that Kirwan and Hutton were slugging it out, there were those who tried to harmonize the Genesis account with the emerging ideas on the antiquity of the Earth. The Swiss Jean André de Luc (1727-1817) for example, believed the six 'days' were allegorical rather than literal, while retaining the full significance of Noah's Flood. Needless to say, his compromising views didn't receive serious consideration in revolutionary France, but made a major impact in conservative England, where he spent the second half of his life.

Another Scottish geologist who was to profoundly affect thinking about the natural world was born the same year that his compatriot James Hutton died. His name was Charles Lyell (1797-1875). Lyell was struck by the long time it must take for geological structures to form, such as the high mountains of the Alps, and for surface landscapes to change, based on the slowness of present-day processes. He spoke in 1850 of the inability to understand the time required 'for grinding down the pebbles of a conglomerate 8000 feet in thickness'. As well as opening up a vista of long periods of geological time, Hutton and Lyell therefore put down the fundamentals for the uniformitarian view of slow, gradual change, which was to influence Charles Darwin in his understanding of natural history.

Charles Darwin himself entered the debate on the age of the Earth. His speculation was based on the time calculated for a region in SE England (the Weald) to be eroded from its original dome-like shape into the large valleys of today. His calculation gave 300 million years, but is based on unsubstantiated assumptions about the height and erosion rate of sea-cliffs - 'an abuse of arithmetic' as the then Professor of Geology at Oxford, John Phillips called it, and so it was. Darwin withdrew the work from later editions of his book. It sits as a mere historical footnote in relation to his other work.

In 1856, the German physicist and philosopher Hermann Ludwig Ferdinand von Helmholtz (1821-1894), who also made major contributions to physiology and psychology, expanded his already exceptionally broad field of expertise to the calculation of the age of the Earth. His calculation made in 1854 was based on the argument that the Sun's energy must be supplied by gravitational contraction, rather than by chemical reactions. If chemical reactions were responsible, the Sun would have a maximum life expectancy of just 5000 years. Instead, the Sun's current size and energy output could be explained if it were collapsing under gravity. He worked out that the age of the Sun must be 21 million years. By inference, we can assume that the Earth originated during the period when the solar system was condensing from a cloud of gas and particles, and so was the same age as the Sun.

The physicist Lord Kelvin (1824-1907), or William Thompson to use his ordinary name, was a leading investigator of the age of the Earth. He attempted to work out the age of the Earth in 3 ways: on the basis of the loss of heat by the Sun; by making calculations of tidal friction; and thirdly, and most importantly, by estimating the loss of the Earth's internal heat. Kelvin, therefore, followed in the footsteps of von Helmholtz, and gradually built up his confidence in von Helmholtz's gravitational collapse mechanism for the Sun. He derived estimates of between 20 and 60 million years for the age of the Sun, depending on how he dealt with the Sun's density structure. Kelvin then tackled the age of the Earth in his paper *On the Secular Cooling of the Earth* in 1862. Kelvin's 1862 estimate gave a range of values of 20 to 400 million years, depending on the values given to important physical factors. These estimates are obviously a good deal longer than the 6000 years estimated from Ussher's biblical chronology, yet short enough to counteract the prevailing opinion amongst influential geologists such as Charles Lyell that the age of the Earth was indefinitely long.

The turf war between Kelvin and the geological community is legendary. Kelvin's estimates may have turned out to be too low, but at least they provided a finite number rather than the indefinite periods of the uniformitarianist camp. The turf war between physics and geology therefore had the added spice of the question of whether the Earth was in a state of perpetual existence, as favoured by the geologists of the time, or was evolving (by cooling) from a definite beginning. Kelvin's attempts to calculate the age of the Earth are discussed in more detail below, since they reveal further aspects of the scientific process.

Returning the Ussher's Dublin, Samuel Haughton was a holder of the Chair of Geology and Mineralogy at Trinity College Dublin between 1851 and 1881. In 1878 he suggested that by measuring the thickness of sedimentary rocks covering the crystalline rocks of the Earth's crust, and then assuming a constant and fixed sedimentation rate, it was possible to calculate the age of the Earth. He and followers came up with all sorts of answers, ranging from 3 to 2400 million years because of the entirely erroneous assumptions they had made about sedimentation rates being fixed and constant. We now know from the dating of sedimentary rocks that rates of accumulation of sediment vary tremendously from place to place. In some places, it barely accumulates at all over hundreds of millions of years, whereas in others it accumulates at the rate of a kilometre every few million years. For example, there are 10 kilometres of sediment off the Mississippi delta in the Gulf of Mexico that we know to have been deposited in the last 60 million years. In addition, Haughton was unaware that sediment is involved in the plate tectonic cycling of the Earth, so some disappears during the process of subduction of oceanic plates beneath the continents.

John Joly, another professor at Trinity College Dublin (between 1897 and 1933) made an estimate in 1889 on the basis of a comparison of the sodium content of the oceans with the annual input of the cation in the world's rivers. He got an answer of 90 to 100 million years. Joly didn't realize that the oceans are in a balance between input and concentration since if the balance is exceeded, salts will be precipitated, which lowers the concentration. The famous astronomer Edmund Halley had carried out a similar calculation more than a century earlier in 1715. He assumed that the oceans had become more and more salty over time as rivers provided ions and evaporation concentrated them. The calculation gave the answer of between 80 and 150 million years. The estimates of Halley, Haughton and Joly were very close to Kelvin's, but all were wrong.

Kelvin's Clanger?

The popular understanding of Kelvin's quest for the age of the Earth[15] is that he blundered by making false assumptions, and that the fundamental problem with his estimates based on the time taken for the Earth to cool from its initial molten state was that he had neglected radioactive heat generation. By the beginning of the 20th century, it was known that radioactive decay generates heat[16], and radioactivity in rocks had been discovered through the work at McGill University, Montreal of Ernest Rutherford and Frederick Soddy and of A. Henri Becquerel in France. By then the question of the age of the Earth had lost all of its earlier associations with rival religious and secularist world-pictures. The situation was different a couple of decades earlier when Kelvin was writing.

Charles Darwin's son George was aware of the discovery of radioactivity by the Curies at the beginning of the 20th century, and recollected that Kelvin's calculations had been made with the assumption of no additional sources of heat in the Earth apart from the slow conduction of heat from the hot interior to the cold exterior. Incorporating heating from radioactive decay into calculations for the age of Sun gave a result of 740 million years, implying that the Earth must also be of this age.

It is true that Kelvin's calculation of the time taken for the Earth to cool from an initially molten state made an assumption of no internal heat sources. In other words the Earth had cooled like taking a baked potato out of the oven. His solution for the age of the Earth required knowledge of the geothermal gradient (that is, the rate at which the Earth gets hotter with increasing depth below its surface) or the heat flux (the rate at which heat passes through a unit area of the Earth's surface). At the time (1884-1886), Kelvin did not have precise information on geothermal gradient and surface heat flux, but based his calculations of the likely range of values. Re-doing Kelvin's calculation based on what we know today about surface heat flux reduces the range of estimates of the age of the Earth to 24 to 96 million years[17].

In the period 1890-1895 one of Kelvin's former assistants, John Perry, questioned the old master's assumptions. He wrote suggesting a reason for the mismatch between Kelvin's calculation of the age of the Earth and the very much longer age anticipated by the new uniformitarianists of geology. He argued that the deep interior of the Earth may transmit heat much more readily than rocks near the surface, which would increase the cooling time calculations of Kelvin by over 50 times.

Perry even suggested that the ability of the Earth to rapidly transmit heat might be at least partly due to the presence of a deep circulation, or convection. It is common knowledge that stirring a sauce on the hotplate allows it to release heat much more rapidly than leaving it unstirred. If the assumption were made that convection allowed a rate of transfer of heat many times the rate possible with conduction, the calculated age of the Earth would be greater than 1 billion years.

John Perry's arguments about the possibility of convection were largely ignored, which was a missed opportunity to grasp a key factor that would, in time, help explain the mobility required for continent drift. In 1881 the Reverend Osmond Fisher published *The Physics of the Earth's Crust* [18], in which he argued that the crust rested on a weak substrate that was capable of flow. The existence of a deformable foundation for the Earth's crust meant that the continents could 'float' and potentially move laterally. Fisher's book of 1881 and Perry's letters in the journal *Nature* in 1895 were published when Alfred Wegener was a young boy, and several decades before the idea of large-scale convection was embraced and popularized by Arthur Holmes.

Today, there are hundreds of journals publishing the results of geological research, ranging from general, multidisciplinary journals such as *Nature* (Macmillan, London) and *Science* (American Association for the Advancement of Science), to highly specialist journals publishing papers is a narrow sub-discipline of geology. It is impossible to read them all, and even those who try to scan new issues of journals for content of interest to them are likely to miss a lot. Back in 1895, there were far fewer journals, and the same argument does not apply. Consequently, there must be social factors that caused these nascent ideas of convection to be forgotten for decades.

Professor of geophysics at Oxford University Philip England[19] suggests that rhetoric was used instead of scientific argument, and that Kelvin and Perry were not understood by a geological community averse to the use of quantitative physical science. More deeply, Lord Kelvin was a mathematical physicist who believed that simple approaches were required to solve complex problems, whereas geologists of the day were engaged in an essentially inductive exercise of gathering disparate observations together in the belief that the 'truth' would emerge. They therefore held a different world-picture – one that made them distrust the oversimplified theories of the physicists – a situation that sadly persists to the present day.

But was Kelvin's clanger that he had neglected radioactive heat generation? Incorporating radiogenic heat production into Kelvin's calculations in fact makes little difference to the solution for the age of the Earth[20]. More important, as John Perry realized, was the ability of the interior of the Earth to transmit heat by convection.

Would you not expect an extremely gifted and experienced physicist who had made fundamental contributions to the understanding of electricity, thermodynamics and heat, and who had patented new instruments such as the galvanometer and the mariner's compass, would graciously and perhaps even excitedly accept the new findings of convection and radioactivity? Yet on the basis of admittedly anecdotal evidence, he did not. There may be many reasons for this. Kelvin thought he had sound evidence that the Earth was solid. For example, it is well known that the gravitational attraction of the Sun and Moon causes tides in the ocean that cause water to rise and fall at the coast. That same gravitational attraction also affects the solid Earth. From measurements of so-called Earth tides, Kelvin believed that the bulk of the interior of the Earth was rigid, and, he thought, incapable of flow. Secondly, the relatively young age for the Earth he calculated was backed up by calculations based on the loss of heat of the Sun, which bolstered his confidence that the result was correct. He was doubtful of the slow rates of evolution implied by Darwin in his *Origin of Species* (1859) since he believed that the Earth was once molten and had cooled down to allow life to develop relatively recently. He was thus drawn into conflict with geologists and evolutionists including T.H. Huxley. There's nothing quite like a disagreement to stimulate further work, and the rudeness and intransigence shown by Kelvin and his supporters towards geologists in general, and to Hutton's work in particular, certainly did the trick. I doubt that the flow of rhetoric and intransigence was in one direction only.

Max Planck (1858-1947), the German physicist who originated quantum theory, remarked on the difficulty that old physicists have with accepting new discoveries:

'A new scientific truth does not triumph by convincing its opponents and making them see the light, but rather because its opponents eventually die, and a new generation grow up that is familiar with it'.

It has been suggested, perhaps overly harshly, that Kelvin's difficulty in coming to terms with convection, radioactivity and the antiquity of the Earth put him into this category of 'old physicist'[21].

The Radiometric Clock

By the last decade of the nineteenth century, there was intense activity and debate aimed at trying to establish the age of the Earth. The discovery of radioactive decay provided a direct means of dating Earth materials, and therefore of working out the age of the Earth.

Certain isotopes, for example ^{238}U (uranium 238) spontaneously decay to produce daughter products such as ^{206}Pb (lead 206). There is nothing we can do to prevent this, which is why the safe disposal of radioactive waste is such a problem, and it occurs at a constant rate. It is known that after 4500 million years, half of the original atoms of ^{238}U will have decayed to the daughter product ^{206}Pb. The 'half-life' for ^{238}U-^{206}Pb is therefore 4500 million years. This forms the basis for calculating the age of a rock containing ^{238}U and ^{206}Pb. Another reaction involving radiogenic decay is the formation of ^{87}Sr (strontium 87) from ^{87}Rb (Rubidium 87). When these techniques were used, pioneered by Arthur Holmes[22], the ages of some rocks were seen to be very large – thousands of millions of years. Holmes' estimate of the age of the Earth in 1927 of 3 billion years was refined to 4500±100 million years in the 1940s, very close to present-day estimates. The race was on to find the Earth's oldest rock. Four billion years is about the oldest rock dated so far. This is not the age of the Earth, but it means that the Earth cannot be younger than 4 billion years.

In 1926 the National Academy of Science of the USA adopted a radiometric time scale, after which the estimates of the age of the Earth have changed little, subsequent minor adjustments having been made possible by improvement in isotopic techniques and in the technical specifications of the spectrometers that generate the numbers.

Radiometric dating has become increasingly sophisticated and accurate, and extends to a very wide range of isotopic systems. This is not to say that geologists do not argue about radiometric dates a lot, or that the dating methods cannot be improved. However, from a scientific perspective, it is inconceivable that a rock with a radiometric age of hundreds or thousands of millions of years could in fact be as young as a few thousand years, as required by 'young-Earth' protagonists.

Meteorites are lumps of rock that were not incorporated into planets when they formed by gravitational attraction of pieces of dust and rock in a cloud of extraterrestrial debris. Radiometric dating shows them to be as old at 4566 million years. Radiometric dating of lunar rock recovered by the Apollo missions also gives an age of about 4500 million (4.5 billion) years. Assuming that the Earth accreted at the same time as the smaller lumps of rock we see as meteorites, the Earth must have formed some 4.5 or 4.6 billion years ago. 4.54±0.05 billion years is the currently accepted value, derived from the Canyon Diablo meteorite[23], which produced Barringer Crater, Arizona, USA.

So after over 350 years of trying, the age of the Earth is known. Plotting estimates of the age of the Earth against the date of the estimate, there is a steady increase in the calculated age over time. The increasing trend, however, flattens out, reaching a plateau at about 4.5 to 4.6 millions of years. Early estimates were, let's face it, profoundly wrong, but over time they became less wrong. The lesson, apparently, is that if you study a problem long enough you will converge on to a reliable answer. But how long do we have to wait? How do we know that the utterances of scientists at any one point in time are correct?

There now is no doubt as to the huge antiquity of the Earth (4.54 billion years ±1%). To believe in a young Earth and 6 literal days of creation requires a truly vast amount of accumulated knowledge to be ignored and contributes to a 'March of Unreason'. This sort of literalism was never part of early Church orthodoxy yet is alive and kicking today in some quarters. That this is so is an illustration of the very powerful impact of world-pictures.

8 KILLER BLOWS: FIRE AND BRIMSTONE

*In which polarized views compete to explain the end of the dinosaurs,
and questions are raised about the nature of evidence . . . .*

The Kill Curve

I have some sympathy with Oscar Wilde when he said 'Education is an admirable thing, but it is well to remember from time to time that nothing that is worth knowing can be taught'. I am also in some agreement with C.S. Lewis who said 'The task of the modern educator is not to cut down jungles, but to irrigate deserts'. Personally, I have learnt most when I have had to teach it, and I have learnt least when settled into my intellectually safe comfort zone.

In geology, as well as in other sciences, we have a range of topics of outstanding potential interest to students and publics. We also have information that is dull to learn but important to know. Putting together a curriculum in a science such as geology is therefore a challenge of balancing what's indispensable to know in order to be well trained with what can encourage the student to think creatively, independently and critically. If you need an operation, you hope that the person doing it is well trained in surgery rather than in diagnosis. Geologists are asked to be both surgeons in their forensic investigation of rocks, minerals, fossils, sediments and soils, and consultants in the understanding of processes relating to these materials.

As an undergraduate I was not very keen on palaeontology. Practical classes invariably involved placing trays of fossils on a long desk, with the simple barked instructions, 'describe them, give them a name, and learn both'. Palaeontology seemed to me to be an infinitely dull subject involving the study of an infinitely large number of dead things, many of which looked depressingly the same. It was like trying to memorize a very long shopping list but not knowing what recipes were being planned. I was right about the dead things, because 99.98% of known species are dead species, a phrase often repeated but originating from University of Chicago palaeontologist David Raup. However, a dull subject? We need to pause. For hundreds of years the taxonomic study of fossils (meaning the description of them, the naming of them and the classifying of them) provided raw data of inestimable value to the burgeoning field of geology. It might seem tedious to you and me, but this remains a very important part of palaeontology. Without the taxonomy of fossils, all the exciting stuff isn't possible. Or at least the exciting stuff would lose its sure foundation and would become unconstrained conjecture.

Would you employ a micropalaeontologist in the oil industry if they couldn't tell you what bugs had been brought to the surface during drilling operations? Does it matter about esoteric concepts such as punctuated equilibrium or phyletic gradualism (Chapter 3) if you can't identify a fossil lying inertly under your microscope? One of the pitfalls of modern university education is that we are in danger of producing a generation who apparently know the exciting stuff without knowing the fundamentals. However, if you teach fundamentals, you get kids like me who regard it as dull taxonomy rather than problem solving. Yet the field of palaeontology, and its near-synonym palaeobiology, has furnished some of the best researchers, thinkers, writers and communicators of recent times[1].

Happily, I have changed my mind about palaeontology, and the beginning of this change can be linked to when, as the new professor of Geology and Mineralogy at Trinity College Dublin, I was faced with lecturing to brand new students who knew close to nothing about geology. In a course called 'Earth, Wind and Fire', I introduced the idea of mass extinctions, and talked about what is sometimes called the 'kill curve' – the percentage of species going extinct as a function of geological time[2]. There's nothing like a calamity for grabbing the attention of students. The prospect of death on a massive scale adds an attractive frisson to the student mind.

These peaks in extinction rates, or mass extinctions, have been the subjects of much debate about their causes[3]. None of the mass extinctions has been as hotly contested as the Cretaceous-Tertiary (K-T) boundary event 65 million years ago, now termed the Cretaceous-Palaeogene (K-Pg) event, which was responsible for the demise of those (mostly) cold-blooded, favorite creatures of children and paleontologists alike, the dinosaurs. Before considering the arguments about the K-Pg event and the strange course of the debate, we need to place the problem in the broad geological context of the last 543 million years of the history of life, known as the Phanerozoic.

Mass extinctions are defined as instances where there is a rapid development of an excess of extinction rate over speciation rate in macroscopic life. This means, since microbial life is excluded, that the scale of the mass extinction can be readily calculated but the value obtained does not refer to the total diversity of life. Microbial life isn't included, not because it's not important but first, because it's very difficult to assess its diversity since almost none of it is preserved in the geological record, and second, because microbes are the planet's great survivors, and doubtless have not fallen prey to extinction in quite the same way as their more exciting macroscopic cousins. Of the macrofossils, marine organisms far outweigh land-dwelling types in the calculations of extinction and speciation, simply because the preserved record of marine fossils is superior.

Over the Phanerozoic, there have been five events where over 50% of marine animals have become extinct, though these are supplemented by up to about 20 smaller mass extinctions. There has been a certain amount of statistical reworking of the data, which has made the Big Five less conspicuous, and some mass extinctions have been suggested for either glorious promotion to or humiliating demotion from the Big Five[4]. Nevertheless, the consensus view is that the mass extinctions are real, and need to be explained.

A number of questions immediately arise. What caused the mass extinctions? Was there one cause for all, a different cause for each of the

Big Five, or multiple causes for each event?

Starting with the oldest, we have the Ordovician-Silurian extinction event (450-440 million years ago) during which 60-70% of species and 57% of all genera became extinct in two main events. The Late Devonian extinction (375-360 million years ago) eradicated 70% of species and 50% of all genera over a period of about 20 million years. The extinction event that took place at the end of the Permian period (251 million years ago) killed off a staggering 90-96% of species and 83% of all genera. It is known as the Great Dying. A prolonged time in the Permian was characterized by high extinction rates for marine life, well before the Great Dying, and there was a long period for recovery, lasting 30 million years for marine invertebrates. The Triassic-Jurassic extinction event at the end of the Triassic period (200 million years ago) caused 70-75% of all species and 48% of all genera to become extinct, leaving land-based dinosaurs in a dominant position. Finally, the Cretaceous-Palaeogene extinction event (65.5 million years ago) resulted in 75% of all species and 50% of all genera becoming extinct. All dinosaurs unable to fly became extinct, and birds and mammals took over as the dominant land-based vertebrates.

Darwin thought that the struggle of species for food and space was more important than external physical causes for extinction. He believed that the variability of the record of extinction was more due to the vagaries of the fossil record in sedimentary successions with many gaps of unrecorded time. However, for the Big Five, most attention has focused on the effects of external physical phenomena ranging from massive volcanic eruptions, impacts by extraterrestrial objects, changes of sealevel, rapid changes of climate and de-oxygenation of the oceans. Each of these mechanisms should leave a fingerprint of its occurrence in rocks, sediments, and the fossil record itself. Each fingerprint should be associated with a certain pattern of extinction of different animal and plant groups living in a range of different habitats, with a certain time scale of extinction and recovery, and with a certain geographic spread.

When this reasoning is applied to the main extinction events of the Phanerozoic, meteorite impact is a contender for just one of the Big Five – the K-Pg boundary event[5] . Extensive volcanic activity is linked to several mass extinctions, including the end-Permian Great Dying. Perhaps surprisingly, the de-oxygenation of the oceans, which takes place mostly during global sealevel rises, is implicated is almost all of the mass extinctions studied. This is not the most glamorous and breathtaking cause of mass extinctions of life, but appears to be both common and effective.

A Cameo of a Controversy

Some palaeontologists, notably David Raup and Jake Sepkoski of the University of Chicago, the originators of the 'kill curve', suggest that variations in extinction rate and biological diversity occur in long-term cycles. Taking extinction rate, they calculated a periodicity at 26 million years based on 12 important extinctions over the last 250 million years. If periodicity were present, it would have implications for whether some internal, biotic organization that led to enhanced extinction in evolutionary 'bottlenecks' had taken place, or alternatively would point to recurring external physical causes. The generation of periodicity by internal organization can be illustrated by imagining the dropping of sand grains onto a pile, as in an egg timer, which grows more and more until the pile collapses, only to build again. Avalanches of sand are periodic, despite the feed rate of sand from the egg timer being constant. In an analogous fashion, apparently periodic mass extinctions may result from simply the way that life is organized on the planet. Alternatively, periodic external mechanisms may be celestial, such as repeatedly crossing the plane of the galaxy through busy lanes of comet traffic, as an example. The course of the debate over the presence of periodicity in extinction rate is a cameo of the scientific process in action, and shows the first warning signs of pathological symptoms[6], as we shall see.

The year before David Raup and Jake Sepkoski published their paper in *Proceedings of the National Academy of Sciences* in the February issue of 1984[7] suggesting a 26 million year periodicity in extinction rate, Sepkoski had presented the results at a symposium on 'Dynamics of Extinction' in Flagstaff, Arizona. Books of Abstracts are circulated at scientific meetings, and the Flagstaff meeting was no exception. An abstract summarizing the oral presentation was therefore available to all those attending the symposium. It is also normal for science writers, bloggers and editors to attend conferences, where they gain a sense of what is controversial, new, and liable to be of interest to their readers. Raup and Sepkoski's extinction analysis attracted their attention, and pieces were published in the science press shortly afterward the symposium in 1983[8]. This broadened the awareness of the model of periodic extinction, and a number of physicists, astrophysicists and geologists requested reprints from the authors. These were dispatched in October 1983, just about when the article destined for publication in *Proceedings of the National Academy of Sciences* was ready for submission.

In the 19 April 1984 issue of *Nature*, a bundle of five articles were published, all involving astrophysicists making use of the extinction rate analysis as a platform for their investigations. Different interpretations were given in those five papers, but the common theme was the astronomical reasons for periodicity in the impacts by comets on the Earth, which in turn were held responsible for mass extinctions. The intriguing question is whether the circulation of pre-prints before publication amounted to a selective disclosure to an inner circle of scientists who were positively disposed to uncritically believe the extinction analysis. In doing so, others who were not part of the inner circle and who might be hostile to the extinction rate analysis would be excluded, thereby biasing the scientific process.

The British stratigrapher and palaeontologist Tony Hallam wrote a 'News and Views' article in the 19 April 1984 issue of *Nature* (p.686, vol. 308, 19 April 1984) in which he critically evaluated the palaeontological basis for the periodicity that Raup and Sepkoski claimed to have found. In the same issue (p.685), the chief editor of *Nature*, John Maddox, expressed concern at the wide circulation of pre-prints prior to publication, which ran the risk of being discriminatory against those not on the mailing list of the authors.

Without any intention by individuals to distort the scientific process, a group of papers in the world's foremost general science periodical had been submitted with an uncritical acceptance of the central tenet of a paper acting as a source that itself had not yet been published. This prevented potentially hostile papers from being included in the set of papers in *Nature*, and thereby sent the scientific debate spinning off in a less than balanced direction. Things have changed a little since the 1980s.

Nowadays, there are embargoes on releasing new research results accepted for publication to the press. However, the continuing potential dangers of scientific 'clubbiness' are apparent.

A further cause for concern is what is known as the peer-review system – the vetting of the quality of scientific work by other scientists, which should be as rigorous and objective as possible. Results submitted for publication are generally seen by two or three reviewers, who comment on the suitability of the work for publication, and commonly provide a great deal of useful input that strengthens the paper. However, through this peer-review system, scientific results are in danger of being released into laboratories, teams of collaborators, or even more widely into the scientific community, prior to publication. In some cases, a reviewer may gain more that simply the arousing of his or her intellectual curiosity. A competing laboratory may benefit from the new insights gained in the review process, and some may benefit commercially. Journal editors now require reviewers to declare that they have no conflicts of interest, but it is a grey area.

It is a source of frustration by scientists that the peer review process sometimes appears to be something of a lottery. Success or failure to publish in the journal of one's choice often appears, especially to the more sceptical scientist, to be at the whim of reviewers who seem lop-sided in their outlook, confused and inexpert. It is indeed an imperfect, human system, but one that is difficult to improve upon.

There is therefore far more to the scientific enterprise than meets the eye. It is far from clinically precise. Tony Hallam summarizes the debate of periodicity in the extinction rate by writing[9] 'Nowadays there is little talk in the palaeobiologial community of a periodicity in mass-extinction events'. The Raup and Sepkosky paper that started it all had 469 citations by the end of 2013, so would be called a citation classic, yet it is not accepted by the community that is best placed to judge it[10]. The presence of periodicity in the record of extinction continues to be a contentious issue.

Mass extinctions may not be periodic, but the fossil record suggests episodic major upheavals in life quite unlike the gradual unfolding imagined by geologist Charles Lyell and naturalist Charles Darwin. The history of life on Earth is more like the journey in the *Rime of the Ancient Mariner* [11] – long periods beset by calm 'as idle as a painted ship upon a painted ocean' and short bursts of breathless activity when 'the ship drove fast'. We could now depart on an excursion into the historical background for ideas of gradual change (uniformitarianism) *versus* abrupt change (catastrophism)[12], but this excursion follows a well-worn path. Instead, we focus on the two major competing theories for the K-Pg mass extinction that resulted in the loss of the giant reptiles that had enjoyed such dominance for the preceding 200 million years.

Father, Son and Scaglia Rossa

The greatest sign of a restless Earth in England is the tinkling of teacups suspended from little hooks in old-fashioned kitchens. Set against this background, no wonder the general public is terrified by the prospect of earthquakes being unleashed during hydraulic fracturing operations to extract gas from shale. The same is not true of the Mediterranean, caught between the northwardly mobile Africa and the stubborn mass of Europe-Asia. On the scale of natural hazards, Italy, the crucible of much of Western civilization, ranks very high.

About 30 million years ago, a mere blink of the eye in geological terms, the terrain making up present-day Italy swung anticlockwise from its position adjacent to southern France, leaving in its wake a new ocean basin and a couple of fragments that got left behind in the form of Corsica and Sardinia. Italy continues to rotate, so that at some stage in the future Italy will join Croatia, Albania and Greece, and the Adriatic Sea will be a distant memory. The movement of Italy means that it crumples at the front, forming great folds a kilometre in height, spaced regularly like lines of infantry. The older, inactive folds are seen on land in the province of Marche but young folds continue out to sea, where they are slowly growing beneath the waves. Each increment of crumpling of the Italian crust folds it like a rucked up tablecloth.

Driving inland from Marche to Umbria, the landscape changes, and the train of equally spaced folds changes to a more open countryside of wide, flat-bottomed valleys separated by ridges on which ancient citadels such as Perugia perch. In this region the crust is stretching apart, as the front of the Italian peninsula pulls away from the rear, like the movement of an earthworm. The stretching produces particularly damaging earthquakes. Recent earthquakes destroyed a large number of frescoes by Giotto in the cathedral at Assisi in 1997, and another hit the town of l'Aquila in 2009, with much loss of life. The stretching apart also allows melted rock to ascend to the surface in volcanoes, which are scattered from Sicily, through the Bay of Naples, and into the blue expanse of the Tyrrhenian Sea. Situated at this change from crumpling to extension sits the historic Umbrian town of Gubbio.

Gubbio is situated at the foot of a fault plane that causes a drop of topography into a flat, cultivated plain. Above the plain, deep valleys have incised into sedimentary rocks initially deposited in deep water as chalky oozes, but which are now folded, compressed and fractured, and visible in scars in the wooded hillsides above Gubbio. At one of many bends in the road descending to Gubbio is an unremarkable-looking exposure of these rocks, with a small notice announcing that here lies the Cretaceous-Tertiary boundary. This modest exposure of rocks provided the observations that sparked an intense debate.

Luis Alvarez (1911-1988) was a Nobel laureate physicist at Berkeley when his geologist son Walter began working in central Italy. Walter had taken a PhD from Princeton in 1967 and was following up his side interest of archaeological geology in Italy. While at Lamont-Doherty Observatory, Columbia University, New York, he started a research programme on the tectonics of the Mediterranean and studied the reversals of magnetic field found in Italian rocks. In 1977 he showed his father a piece of rock made of a pale-coloured limestone and a pink limestone separated by a dark clay layer from the Scaglia Rossa rock formation (literally 'red scaly', which refers to both its generally pink colour and its thin layering, which causes it to fracture and break apart like scales) near the town of Gubbio. The dark clay layer marked the boundary between the Cretaceous and the oldest part of the Tertiary period, the Palaeogene. The locality has since been so heavily visited and sampled that the dark clay layer between the underlying Cretaceous and overlying Palaeogene sedimentary rocks is hidden in a deep cleft as long as your arm.

Father and son pondered on the length of time represented by the dark clay layer. With his background in nuclear physics, Luis contacted colleagues who could measure very low concentrations by neutron activation. They thought that the platinum group of elements, of which iridium is one, would be suitable because these elements are supplied to the Earth in a very gentle cosmic rain of micrometeorites. To their surprise, levels of iridium were hundreds of times the level expected as background. Eventually, they identified the culprit as a meteorite that must have struck Earth at the time of deposition of the Scaglia Rossa. Working with colleagues who were expert in rock magnetism, they came to the conclusion that deposition of the clay layer represented just 10,000 years of time.

The hypothesis published in the prestigious journal *Science* in 1980[13] proposed an asteroid impact that had caused a toxic 'nuclear winter', the effects of which cascaded through the food chain to cause extinctions of animals and plants. There was excitement at this discovery, but also widespread opposition, particularly from palaeontologists, whom Luis Alvarez disparagingly called 'stamp collectors', a tiresome form of put-down started by Lord Kelvin and continued by Ernest Rutherford for people who were troublesome in disagreeing with them. The social factors of the debate, in this case a turf war, are at this stage starting to emerge.

Things began to stack up in favour of the impact hypothesis in the years after publication of the *Science* paper. The likelihood of impacts by rocky asteroids and the faster but lighter dirty balls of ice known as comets became better understood and accepted. Iridium anomalies, as seen as Gubbio, were found at over 100 other sites scattered around the world. Tiny spherules representing airborne droplets of melted rock were found at several K-Pg boundary localities, and quartz crystals with features characteristic of having undergone a high-pressure shock were found. Crystals of spinel with extremely high nickel contents typical of meteorites were found in a layer much more concentrated than the distribution of iridium. High abundances of soot and charcoal were found in the same layers as the iridium, suggesting widespread fires at the K-Pg boundary. All this fitted well with the impact hypothesis. However, a contender for the K-Pg mass extinction was shortly to arrive on the scene.

India's Smoking Gun

The Deccan is a large region of central-west India underlain by a thick (< 2 km) pile of basaltic lavas, most of which were erupted onto a land surface, whereas some flowed down a topographic slope to the southeast and entered an embayment of the Bay of Bengal. Since their emplacement, the western flank of India has been uplifted like the edge of a stale slice of bread, and rivers have cut deep valleys into the lavas. The hillsides are fretted like a staircase due to the different hardness of the lavas and intervening sedimentary rocks, which explains the name of 'Deccan Traps', since a trap is a stair step from Nordic languages.

At the beginning of the 1980s, the exact age of the volcanic eruptions responsible for the Deccan Traps wasn't known. However, studies of rock magnetism in the Deccan area of India using newly developed techniques showed that the entire pile of layered basalts had been formed from volcanic eruptions that occurred within a period of time containing just two magnetic reversals. The reversed period during which the main eruptive phase (Phase 2) took place was chron 29[14]. This is exactly the same time zone as that containing the iridium-rich clay at Gubbio. There was therefore clear evidence that the eruption of the Deccan basalts and an extraterrestrial impact took place at the same time. The case was clinched in the 1990s when a joint French-Indian team analyzed the lavas from the Kutch region of India. They found an iridium-rich layer *within* lavas belonging to chron 29R. This was proof that whatever produced the iridium, which most scientists believe was an impact of a meteorite, took place while the volcanoes in India were erupting. Strange and coincidental as it may seem, the only sane conclusion is that the two events happened at the same time. The uncertainty is how each contributed to the mass extinction that wiped out the dinosaurs.

Now there were two opposing ideas, one involving extraterrestrial impact, the other involving massive volcanic eruptions. The volcanic hypothesis was quickly developed, and downgraded or incorporated many of the observations previously used to support the impact hypothesis. High levels of iridium were explained by the sheer scale of the volcanic eruptions, which tapped deep levels in the interior of the Earth. The tiny spherules were attributed to volcanic processes. Only the shocked quartz and nickel-rich spinels proved difficult to account for. The large volcanic eruptions were presented as a new means of explaining the biotic crisis that caused the K-Pg mass extinction.

A number of questions were being asked. Was there one major, sudden extinction that would support the impact theory, or was there a step-wise, more prolonged pattern of extinction that would favour the volcanic hypothesis? If a large volcanic eruption had played a vital role in the K-Pg mass extinction, then what about the other four mass extinctions making up the Big Five? Such eruptions are common in Earth history, so could it be entertained that they have been responsible for all the major mass extinctions?

The results of rock magnetism studies on the Deccan basalts showed that they had cooled down from lavas erupted at a latitude of about 30°S. They had obviously been transported northwards as passengers on the Indian subcontinent since their formation 65 million years ago. The island of Réunion has been volcanically active over the last couple of million years, and across the sea floor of the Indian Ocean are chains of volcanoes that head like an umbilical chord to the Deccan. The conventional view is that the Deccan formed from volcanic outpourings over a 'hot spot' 65 million years ago, and by drifting northwards has left a trail of progressively younger volcanoes leading towards its present position beneath Réunion.

There are many other hot spots. The largest today is under Hawaii in the Pacific Ocean. Iceland is another located in the ocean, and Yellowstone is an example situated on a continental plate. These are currently or very recently active, but delving into the geological past, there are examples that look very similar to the Deccan. The Ontong-Java Plateau of the western Pacific is built of basalts that are 110 million years old. The Parana Traps of South America are 133 million years old, and are the same age as the Etendeka basalts of Namibia. The Karoo lavas of southern Africa are 184 millions years old, and the largest of them all, the Siberian Traps of northern Russia are 250 million years old.

Comparing these periods of major eruptions giving rise to Large Igneous Provinces with times of mass extinctions yields some good fits, but also some ambiguity. Are the good fits simply a coincidence?[15]. The best fit is for the Siberian Traps, which formed at the same time as the Great Dying at 250 million years ago. Rock magnetism data suggest, as in the Deccan, that the 3-km stack of basalts was emplaced within one magnetic zone reversal, so the duration of the eruptions was short. There are no elevated levels of iridium, no spherules and no shocked quartz. So there is a close correspondence between massive volcanic activity and extinction, but no connection with an impact in the case of the Great Dying.

The results of other comparisons are more ambiguous, and one sees the fit as good or bad depending on one's standpoint. Vincent Courtillot was optimistic enough about the correlations to write a paper in 1994 with the title 'Mass extinctions: seven traps and one impact?'[16]. That one impact is at the K-Pg boundary, but Courtillot favours the Deccan volcanic eruptions as the smoking gun for the associated mass extinction. According to him, the volcanic trigger seems to have been released fairly periodically, with a typical interval of 30 million years.

What was needed was better palaeontological and sedimentological evidence of the precise nature of the K-Pg mass extinction, and some modelling of what might have happened. Both of these initiatives took place, but the debate became a lot more interesting with the discovery of a prime candidate for an impact site in central America.

The Crater of Doom

The presence of shocked quartz in rocks at the K-Pg boundary[16] suggested that if an impact happened it must have hit continental crust, which underlies the land surface and shallow continental shelves. Quartz is absent in the oceanic crust. In addition, the abundance and size trends of the shocked quartz crystals suggested that the impact site must be close to North America. If the impacting body had hit the sea, a tsunami would be generated that should leave a trace of its passage. In the early 1990s, occurrences of relatively thick (metres to tens of metres) deposits in Cuba, Haiti, Mexico, and from deep sea drilling sites in the southwestern Gulf of Mexico, all looked to be candidates for tsunami deposits washed in from an impact site, containing spherules and shocked quartz.

At about the same time, a little known abstract by two petroleum geologists published a decade earlier came to light. This abstract described a buried circular, 200 km-wide structure on the northwestern edge of Yucatán that, although invisible at the surface, could be recognized from gravity and magnetic surveying. Close to the centre of the ring-like structure is the small town of Chicxulub. Drilling close to the village revealed lavas, ashes, broken-up debris and shocked quartz, all of which were reminiscent of the filling of a crater after an explosive impact. Dating of volcanic glass recovered from the drill core provided a very precise age[17] of almost exactly 65 millions of years ago. The same age was obtained from volcanic glass found in nearby Haiti. It was looking increasingly as if the Crater of Doom[18] had been found.

New seismic investigations[19] of the Chicxulub crater in 1996 and again in 2005 showed that the impact structure was 180-200 km in diameter, and rocks at least 30 km deep were disturbed. These observations not only confirmed that the Chicxulub structure was indeed an impact crater, but also provided raw data for computer modelling of the impact process[20].

A Small Question of 300,000 Years

Paleontologists have never much liked the impact theory. In fact, they have never been keen on grand theorizing about mass extinctions in general. Palaeontologist and stratigrapher Tony Hallam from Birmingham University, echoing pleas for inductive reasoning that we have stumbled over elsewhere in this book, wrote scathingly in *Catastrophes and Lesser Calamities* (2004, page 158):

> ' Before astronomers indulge in further speculations about the cause of mass extinctions they would do well to learn something about the rich stratigraphic record of their own planet',

adding that

> 'too many theoreticians were chasing too few facts, a situation very different from geology'.

This suspicion of grand theories for mass extinctions may have developed for various reasons, the foremost of which is that the details of the patterns of extinction at the K-Pg boundary are complex and do not, in the view of palaeontologists, unambiguously support the idea of a sudden, once-only catastrophe. It boils down to whether the dinosaurs and other species died out 'not with a bang but with a whimper'[21]. The crusade against the impact theory led chiefly by a section of palaeontologists has put the controversy once again on the front pages.

The most vocal of those opposing the extraterrestrial impact theory for the K-Pg mass extinction is Princeton University micropalaeontologist Gerta Keller. Keller has studied the K-Pg boundary in a number of localities, but has focused on exposures along the Brazos River in Texas, where she believes the evidence is conclusive. She found a thin clay layer with spherules that she attributed to the Chicxulub impact, but proposed that its deposition was significantly earlier (by 300,000 years)[22] than a second clay layer with high iridium concentrations found above it. In between these distinctive clay-rich layers are several metres of sediments that were, it was claimed, deposited slowly. This requires two impacts, not one – the first responsible for the spherules (Chicxulub), and the second responsible for the iridium, some 300,000 years later at the K-Pg boundary, coinciding with the mass extinction. If Keller were right, the connection between the meteorite impact and the mass extinction would loosen fatally.

In an attempt to reassert the Chicxulub impact as the cause of the K-Pg extinction, 41 authors joined in writing a review in the journal *Science* in 2010[23]. The paper marshals the arguments in favour of the impact hypothesis. One thing is certain: the distribution of ejecta based on 350 K-Pg sites around the world shows very clear proximal to distal (close to far-away) changes that uniquely identify Chicxulub as the impact site.

The debate currently rests with the interpretation of the sediments between the spherule layer and the iridium layer in proximal locations such as Brazos River, Texas. In more far-away locations, there are no intervening sediments, and the iridium layer has provided perfect ages of 65.5 million years. This lends support to the view that the intervening sediments are reworked and transported following the Chicxulub impact, and were followed by the late-stage deposition of iridium-rich clay. If correct, the problem of the alleged 300,000 years mismatch would fall away.

Science by Media

As in other controversies, the direction of the scientific debate is far from linear, and frequently bad-tempered. Social factors have been at work in the evolution of the K-Pg mass extinction debate. An important aspect of the debate over the last several years is the role of the media.

Any research result that throws light on the extinction of the dinosaurs tends to cause a media frenzy. Almost anything said or written will be lapped up by the media without much interest in its scientific basis. Media exposure of research is a part of the academic's outreach, and is generally a good thing, especially bearing in mind that most universities consume tax-payers' money. It is also understandable that the media like a controversy. However, scientific arguments should not be rehearsed there. You don't win scientific points by asserting your views in blogs and digital news sites. The danger is that doing this can become part of a strategy to enhance status in the face of dodgy science, to lobby for a certain viewpoint, to keep the pot boiling.

Although Gerta Keller and co-workers have published several peer-reviewed papers in good journals to support their case, the level of engagement with the media (The Dissenter, Public Lives, Princeton Weekly Bulletin, ABC News, National Geographic News, Geoscientist, Earthsky) is quite remarkable. Without peer or editorial review, these news pieces, blogs and magazine articles always portray Keller and co-workers as the oppressed minority, bravely taking on the unlistening majority. This unlistening majority is itself portrayed as comprising powerful scientists who made their names in developing the impact theory for the K-Pg mass extinction. Gerta Keller said in National Geographic News (Thursday October 28, 2010):

'This is more religion than anything else. A lot of people are so wedded to the [Chicxulub] theory that it seems that no evidence can ever convince them of anything else'.

By 'religion' the author means that those in favour of the Chicxulub impact as the overriding cause of the K-Pg mass extinction have a blind faith in their hypothesis. Yet the same could be said of the adherents of the volcanic hypothesis. Who has priority in terms of evidence, and exactly what *is* the 'evidence'? Each side of the argument must take a position on the interpretation of observations that are themselves contested. And that position will be affected by the prior investment of each side in the hypothesis being advanced.

Charles Darwin wrote in *The Descent of Man* in 1871 (chapter 21):

'false facts are highly injurious to the progress of science, for they often endure long; but false views, if supported by some evidence, do little harm, for everyone takes a salutary pleasure in proving their falseness'.

This confuses matters further, so let's step back and think about the language being used. A fact is a statement of a physical reality, which one might tentatively call a 'truth'. All observers agree with the physical reality, such as 'it rained in Oxford last Wednesday'. Carefully set up and monitored experiments frequently yield facts. For instance, 'the 1 cm-diameter ballbearing with a density of 4000 kg m^{-3} took 4.36 seconds to fall through the 1 m-high column of glycerol that was at a temperature of 21°C'. If this experiment were to be carried out a thousand times (on planet Earth) the same result would be obtained, so the settling time of the ballbearing is a fact. An observation is the noting of a phenomenon by an individual, but is a perception without objective truth. Some people are better at observing than others, and improved technologies commonly allow better observations. Evidence is an observation that is pertinent to a problem under scrutiny. An interpretation adds meaning to an observation or a collection of observations, or is used to make sense of a number of facts.

So we should beware of claims to possess 'evidence' in support of a viewpoint. In the geological sciences, we have a triad of evidence that forms the basis for ideas. We have physical experiments, models, and direct observations, the latter sometimes unwisely being called 'ground-truth'. Different problems require different balances of these lines of enquiry. Perhaps the most interesting is the situation where physical experiments suggest that a phenomenon should occur, and computer models likewise predict its occurrence, but where the observational evidence is lacking or ambiguous. Then, on occasions, and to everyone's delight, new observations are made that support the predictions. Particle physics and the recent discovery of the Higgs boson seems to be a bit like this. In other situations, observations may persistently and awkwardly not 'fit' with conventional thinking. Spotting anomalies may lead to the abandonment of older models and the development of improved ones in the finest tradition of Thomas Kuhn[24]. In the scientific and social cauldron of research, the triad of physical experiment, models and observational evidence simmer and mingle.

Direct observational evidence isn't 'ground truth', because the observations are themselves made with formidable limitations. Measurements aren't much use if they are not connected with strands of meaning. If I want to know the rate that the Earth's surface is eroding, I am challenged to know where I should make my measurements (everywhere?). If I were able to make those measurements, over what time scale would they be applicable? Yesterday, when the river flowed blue and calm between its reeded banks? Or tomorrow when turbulent brown sludge is charging downstream and the floodplain is awash with human detritus? Which is the more typical? How long do I need to keep measuring in order to know the long-term rate of erosion? So we must beware of numbers and descriptions masquerading as ground truth.

When someone, in this case no less than Charles Darwin, claims that 'false facts are highly injurious to the progress of science', they are incorrect, since a fact is a fact, until such time as it can be shown that the universe operates differently to the way we imagine (as in the difference between Newtonian physics, relativity and quantum mechanics). What is injurious to science is the *treating of observations as facts* by the observer.

Equally, when a scientist says of another that 'no evidence can convince them', then one might ask what is meant by evidence? We hear a lot about evidence-based policy. But evidence is tainted by perception. It's a question of your evidence against my evidence, as in a court of law. My so-called evidence is simply my idea of what my observation means in the light of my pre-judgments.

The heat of the debate over the K-Pg mass extinction simply boils down to this. With different world-pictures, the same observations are explained differently. We have, I think, hit upon one of the greatest sources of disagreement in the geological sciences – the mistaken belief held by the observer that their observations are facts, and that they therefore possess conclusive evidence in support of their hypothesis.

Perception (and statistics) also plays a role in the recognition of patterns such as cycles, and once recognized, in the weight given to them in searching for explanations for periodicity. What do we make of Earth's galactic journey of 225 million years? Is there any relationship with the time scale of diversity of genera, or the period of the supercontinental cycle, or the occurrence of icehouse periods? And what about Raup and Sepkoski's suggestion of a periodicity of roughly 26 million years in the kill curve that we looked at previously in this chapter? Is there a correlation between mass extinctions, large igneous provinces, sea level variation, mountain building and sea-floor spreading? Is there a pulse, or set of pulses, in the Earth system? Yet statistics are notoriously used to support pre-judgments. There are 'lies, damned lies and statistics'[25], as the much-used phrase goes.

Vincent Courtillot provides his own commentary on the social factors explaining the course of the K-Pg mass extinction debate[26]. He identifies two factors that played a role: first, the size of the community actively involved in the research into the problem, and second, the competitiveness of the professional science environment, judged for example, by the pressure to publish and to lead the scientific agenda. It is claimed that both of these factors acted strongly in the USA. Luis Alvarez's career made him well adapted to a competitive environment, and his high status as a physicist swayed geochemists, geophysicists and astrophysicists in favour of the impact hypothesis. Geologists and palaeontologists, however, were less easily convinced and were more concerned to intensify direct observation from the various materials available to them – the sedimentary deposits with their special ingredients, and the fossil record. We have come across this tension between deductive and inductive reasoning, and between catastrophism and uniformitarianism previously.

The discourse between the two sides of the argument became bitter, as it remains today (end 2013). Scientists are used to the heat of battle and probably enjoy it, but unfortunately the disagreements can become personal. This tribalism can lead to key players leaving the field to work on something less bruising. This was the case with Chuck Officer, who signed off with his final thoughts in *The Great Dinosaur Extinction Controversy* [27] in 1996. For him, the party was over, but it has continued rancorously for over another decade and a half.

In the K-Pg extinction debate, as in others used in this book, one sees the impatience to disseminate a new idea as quickly as possible, and to get scientific precedence. This means that ideas are often shaped and propagated on the basis of incomplete data collection and analysis, and certainly on the basis of inadequate peer review. The scientific process, when it is working, subjects ideas to close scrutiny before they are published. Such scrutiny may itself be lop-sided, as journal reviewers are themselves human, but to sidetrack peer review in order to secure a media splash is rarely a good idea. The media have their own standards, and they are not the same as the scholarly standards of science. To engage in a war of words in non-peer reviewed magazines makes things worse.

Courtillot also believes that there was an unbalanced attitude in the media. He writes (page 138, *Evolutionary Catastrophes*) with some frustration:

'Why does the public apparently find an asteroid so much more glamorous than a volcanic eruption?

This is less troubling. It is important in the sense of the need for translation of scientific findings to the publics, but no one is in any doubt that the focus of the media is primarily about sales figures and entertainment, not education. The blogosphere is worthy of more concern. In the current fashion for openness and transparency, a number of blogs have sprung up offering views on many issues of the day. These blogs can easily become the instruments of persuasion of militants, offering intolerant and sometimes vindictive comments on their intellectual adversaries. When scientific direction and public understanding is based on who shouts loudest, we will find no shortage of 'ultras' lining up to take the megaphone.

9 AN OLIVE LEAF AND A DOVE: FLOOD STORIES

In which a flood myth persists in scientific explanation, evidence of megafloods appears in unexpected places and legends of plagues and famines receive a face-lift as examples of the 'smoking gun syndrome' . . . .

Going Back to the Ark

Since life was spawned on planet Earth billions of years ago, it has been something of a roller coaster ride. Somehow, life was never completely snuffed out, as is evident from the fact that you are reading this book. But the fossil record shows a pattern of speciation and extinction throughout time, and several periods of especially high extinction rates, known as 'mass extinctions'. Some argue, not fully convincingly, that we are currently experiencing our very own mass extinction, where the cause *is* known – human activities. A story of annihilation followed by a new start was widely believed, to many symbolically, for centuries before a modern understanding of the fossil record revealed by palaeontology. That annihilation was believed to have wiped out every living thing in the world apart from a few chosen survivors. It is the ancient mystery of Noah's Flood. Dismissed by many as outright myth, there may be an underlying historical authenticity. Throughout pre-human and human history, floods have had a profound impact on the physical, biological and cultural landscape of the Earth.

Stories of massive floods are found in many different early

cultures, even in my home from 1996 to 2001 of Ireland, in the opening of the classic Irish epic *The Tain*. The account of Noah's Flood is found in the book of Genesis of the Judaeo-Christian Bible. Its centrepiece – the occurrence of a flood that covered the Earth – has rippled through thinking about the natural world to the present day. Making sense of the account of Noah's Flood is therefore far more than a question of narrow theological interest[1].

There are few biblical topics that have attracted such disparate views as Noah's Flood. At one extreme, we still have today biblical literalists who seem content to dismiss a great volume of theology, archaeology and natural history, not to mention geology, in defence of their interpretation of a truly global annihilation. At the other extreme, we have scientists and secular writers in general who want to throw the baby out with the bath water by concluding that the entire story is mythological and devoid of any historical authenticity. Derek Ager writes in his book *The New Catastrophism* [2] with regard to Noah's Flood:

> 'I do not think the bible-orientated fundamentalists are worth honouring with an answer to their nonsense. No scientist would be content with one very ancient reference of doubtful authorship.'

It is one thing to hurl abuse at so-called bible-orientated fundamentalists – after all, they're used to it. It's another thing to trivialize flood stories by implying that there is 'one very ancient reference of doubtful authorship', when in fact flood stories, as we shall see, occur in a number of early civilizations. Is Ager implying that the Genesis account should have been peer-reviewed and published in a reputable journal? This kind of approach is hardly an encouragement to those who would like to understand why great flood stories occur in so many widespread ancient cultures and whether they reliably document a single natural event, or several different natural events, of profound significance in human history.

The narrative account of The Flood is fundamentally about the faith of a man named Noah and the patience and forgiveness of God. It is also about a new world order, a new beginning after things have gone badly wrong. It is therefore a note of hope for a world facing catastrophe and perhaps has a contemporary relevance as we tear down forests, pollute rivers, lakes and seas, fill the air with poison, and diminish biodiversity. However, the deluge is simply a backdrop; it is the theatre with its stage and scenery and props, but the play concerns the relationship between God and Man. Nevertheless, there is much to learn from the Biblical account that impinges on the historical authenticity of an ancient major flood at the crossroads of human history.

The flood story is prefaced by a genealogy leading to Noah, reminding us of the interconnectedness of people from generation to generation. The genealogy also emphasizes the linear view of time in the Old Testament of the Judaeo-Christian Bible, with a beginning, with distinct and unrepeatable events, and with the prospect of an end[3]. This linear idea of time is rather different to that of other ancient civilizations such as those of China, India, Babylon, Egypt, Greece, the Mayas, Aztecs and Incas, where the universe seems to oscillate eternally. Such a view encourages fatalism rather than confidence. Stanley Jaki[4] (1974) believes this linear view of time was crucial for the explosion of science in Christian Europe in the 17th century.

In the narrative, we are told that God saw that the wickedness of men had spoilt his world and decided to blot out men and other living creatures – all except for Noah. Noah alone was warned of the impending apocalypse and was instructed to make an ark in which he is to save his family and 'two of every kind' of all living things. This word 'ark' is interesting. The Hebrew is 'teba', meaning a box or chest, and is only also used in the Bible for the basket in which the baby Moses was found amongst the rushes of the River Nile. The illustration of Noah's ark in the Nuremberg Bible of 1483 shows a box-like, rectangular vessel with no hull or curved stern. It looks like just about the most un-seaworthy craft possible to design. We already have a hint here that the account is not to be taken entirely literally.

Noah was obedient and was ready when the floodwaters came. We should not underestimate the magnitude of this obedience. To be told to build a gigantic boat in the middle of a dry field would seem rather bizarre. We are told that 'all the fountains of the deep burst forth and the windows of the heavens were opened', and 'rain fell upon the Earth for forty days and forty nights'. The water covered all the mountains, lasting for 150 days, and killing all living things except those in the ark. Eventually the flood receded, and after seven months the ark was beached on the mountains of Ararat (Urartu in Assyrian, which became an important mountainous kingdom north of Mesopotamia and east of modern Turkey). After ten months the tops of the mountains became visible. Noah released firstly a raven, then a dove, but they found no resting place. A week later a dove returned with a freshly plucked olive leaf, showing that the waters had now subsided enough to leave dry ground. After a short further wait, Noah and family and animals disembarked onto dry land.

Although there are flood stories in many ancient cultures, the most complete account of a flood is from the *Gilgamesh Epic* of Mesopotamia. The similarities with Noah's Flood are striking. The Assyrian version of the *Gilgamesh Epic* was discovered on tablets in excavations in Nineveh in 1845, and in fresh expeditions in 1875-76. The epic has since been found in other translations (Akkadian, Hittite) going back to between 1600 and 1000 BC, but the oldest version comes from the Sumerian civilisation dating back to 2100 BC. Before that, the epic probably existed in oral form.

Mesopotamia, now the modern nation of Iraq, was situated in the great floodplains of the Tigris and Euphrates Rivers, and flood stories are an integral part of Sumerian traditions. The fragments of Sumerian text tell of a supreme god, Enlil, amidst an assembly of lesser gods. The gods decided to wipe out mankind in a Flood, but a king called Ziusdra learnt of the impending disaster and built a huge vessel in which to survive the flood apocalypse. Ziusdra was given eternal life as a reward for having preserved the 'seed of mankind'. The purpose of this Flood story seems to have been political, to strengthen the position of the King in an ordered system and to give him godlike status.

Towards 1800 BC the Babylonian empire succeeded the Sumerian, and the Mesopotamian Flood story was modified and rewritten, most notably in the Atrahasis epic. In this epic, humans provide the labour in place of the lesser gods. However, their population explosion caused so great a noise that it disturbed the gods. The gods set about destroying the humans but were thwarted by the god Enki, who had brought humans into existence in the first place, and a mortal devotee, the King Atrahasis. Finally, the supreme god Enlil decided to send a global flood. Enki's clandestine advice to Atrahasis was to build a huge vessel coated in pitch. Atrahasis filled the boat with possessions, animals and birds. After a tumultuous flood, the boat came to rest on a mountain. Atrahasis eventually sent out a dove, which, finding no resting place, returned to the boat. A swallow, then a raven, were sent out, but they stayed away. The floodwaters having receded, Atrahasis then disembarked from the boat.

How is it that these different cultures share not only the same basic framework, but also similar details in imagery? It must be either that all of these ancient peoples were profoundly influenced by tales of a disastrous flood passed on through oral tradition, or that the account has simply been recycled and modified as it was transferred from Sumerian to Assyrian and Hebrew cultures.

There are many features that make clear the symbolism of the Flood narrative. This was accepted a long time before the rapid growth of science. The second-century Greek philosopher and Christian Origen, for example, who spent the first half of his life in Alexandria, Egypt, stressed that Noah's flood should be taken figuratively rather than literally. Indeed, he believed that taking the account literally was to rob it of its lessons for mankind. Saint Augustine (born 354), a couple of centuries later, warned of the dangers of taking Biblical interpretations literally when conflicting with reason, and reason was based on what one could observe. So far so good, you might think. But Saint Augustine believed that reason, as it was then construed, allowed for a truly global flood that covered even Mount Olympus. This reason was based on the observation that fossil shells, plants and bones were found in rocks exposed in the landscapes around him. How else could these fossils have got there? Consequently the use of reason is no guarantee of correctness. It shifts with the tides of history.

We will shortly fast-forward a millennium to witness the tension between new discoveries in the natural world as science took off as an enterprise and the orthodox views that these discoveries challenged. First, we can conduct a thought experiment about the likelihood that Noah's flood covered even Olympus, that place between Earth and the heavens, the highest mountain in Greece, and the home of the gods.

There is a certain amount of water in what is termed the hydrosphere. Water is found as vapour in the atmosphere, as ice and snow in ice sheets, glaciers and permanently frozen ground, and as liquid water stored in the world's biota, rivers and lakes, below the surface of the Earth as groundwater and of course in the oceans. If we make the startling proposition that all of the water in ice and snow, all of the vapour in the atmosphere, and all of the water stored under the surface of the Earth as groundwater is made available in Noah's flood, and if we distribute this water over the entire surface of the Earth, it would be just 100 m deep.

The Earth has an average elevation of the land surface of 840 m, and Mount Everest is about 10 km above sea level. Clearly, even by making a massive assumption about all of the water of the hydrosphere being added to the oceans, we simply do not currently have enough molecules of water on the planet to inundate even the lowest of hills, let alone the highest mountain peaks[5]. In addition, if even the highest peaks were covered with water, where did it all go when the waters receded? The idea of a global inundation of even the highest mountaintops needs to be put to rest.

The 'Heresy' of Natural Science

The question of a global flood forms a subset within a wider debate with geological discovery at its heart; it is an illustration of the struggle between Biblical literalism and natural science[6]. The idea of a literal global flood was accepted by all but the heretical in the 15th and 16th centuries. The writings of the theological giants Martin Luther (1483-1546) and John Calvin (1506-64) support a literal reading of the first book of the Bible, Genesis, where the account of Noah's flood is found. Outright rejection of biblical teachings was not tolerated, and some, like Giordano Bruno (1548-1600) were burned at the stake for denying Noah's Flood. In the 17th century in Britain a debate was started about the nature of the Flood – violent and destructive *versus* placid, with implications for its ability to refashion the Earth's landscape. These, however, were arguments about points of detail, and during the Renaissance the power of the Church was so great that opinions running counter to orthodox religious beliefs were squashed.

Early geological thoughts, and particularly the discovery of fossil sea shells embedded in rock layer upon layer, caused increasing tension between the orthodox beliefs of the Church and these early discoveries of natural science. Leonardo da Vinci (1452-1519), painter, architect, sculptor and engineer, was sceptical about the ability of Noah's Flood to explain the presence of fossil seashells in Italian mountains. He brilliantly concluded that layers of hard sedimentary rocks containing fossils were initially laid down as soft sediment that had settled on the floor of ancient seas. Consequently, those layers of fossil-bearing sedimentary rock found near the tops of ridges in his native northern Italy must have been formed when sealevels were previously much higher. These inspired ideas, and those of a number of learned men in Italy who were contemporaries of Leonardo, were more or less disregarded, and the subject went into a dark age lasting three centuries. I call it a dark age because much effort was spent in mortal combat between competing speculations within the straitjacket of religious orthodoxy, instead of getting on with observing Nature's record.

140

Following in the footsteps of Leonardo, but nearly two centuries later, the Dane Niels Stensen, known as Steno (1638-86), correctly identified sharks' teeth embedded in rock and formulated the idea of a stratification of sedimentary layers. He nevertheless continued to assume that the biblical chronology was correct, and therefore that all of the sedimentary layers were deposited by a single gigantic flood. That was his world-picture. There thus arose a school or tradition of people who made it their business to explain Noah's Flood in terms that would account for the growing amount of evidence in Nature. They are now known as *diluvialists* – students of the Flood.

As an example, the English protestant clerics Thomas Burnet (1636-1715) and William Whiston (1667-1752) both believed in the occurrence of a global flood that was God's punishment of a sinful world, but they were also both convinced that because God operated through natural processes the Flood could be explained in the terms of natural science. Burnet in his *Sacred Theory of the Earth* (1681) proposed that the Earth was originally 'perfect' with a smooth crust. The Flood introduced chaos and imperfection, causing its surface to be heaped into mountains. Burnet wrote:

'There is nothing in Nature more shapeless and ill-figur'd than an Old Rock or a Mountain'.

The idea therefore developed that the present features of the surface of the Earth were sculpted by the Flood. It was not until the middle of the 18th century that the biblical timescale was starting to be seriously questioned, with the result that the biblical flood was gradually reassessed.

One of the great figures of geology in the 18th century was James Hutton (1726-1797), born and educated in Edinburgh. Hutton worked out by careful observation that modern processes visible today, like the erosion of sea cliffs and the incision of river valleys, could explain the geological features of the past. He suggested that the wearing down of the continents by erosion was compensated by the uplift of the land surface by geological movements. He believed that the present was therefore the key to the past. Interestingly, he took what seems a leap in logic by stating:

'We find no vestige of a beginning – no prospect of an end'.

Although he may have been specifically thinking of the repeated cycles of uplift and erosion, a growing realization of the vastness of geological time did not strictly imply that there was no beginning. It was not an objective statement made in the face of ice-cold scientific facts. It was more a counter-attack to the prevailing religious view that somehow erosion 'could only blur the divine masterpiece' of Creation, and that the Earth was a mere 6000 years old.

Elsewhere in Europe, Abraham Werner (1749-1817) was formulating ideas of a primaeval ocean from which all of the Earth's rocks were formed. This became known as the *Neptunist* theory. Neptune was, of course, the god of the sea.

Hutton's views were opposed on two fronts, by Neptunists and by those who believed them to be in conflict with the Bible. Richard Kirwan, President of the Royal Irish Academy from 1799 to 1819, was one of the latter. Kirwan observed in 1793:

'how fatal the suspicion of the high antiquity of the globe has been to the credit of Mosaic history, and consequently to religion and morality'.

History has not dealt kindly with Kirwan. C.C. Gillespie in *Genesis and Geology* [7] (1951) stated that:

'Kirwan presents…almost too classic an example of the degree to which a perverse conception of natural theology could corrupt a scientific mind'.

Kirwan's efforts to discredit Hutton were a last gasp, rather poor attempt to re-assert the priority of the biblical chronology. A later challenge to Hutton's views came from those who believed that the history of the Earth was marked by sudden catastrophes. The Swiss/French anatomist and palaeontologist Georges Cuvier (1769-1832)[8] believed that the Earth had had a relatively tranquil existence interrupted by 'révolutions' or 'catastrophes'. These 'révolutions' were responsible for the extinctions of fossils that Cuvier was able to observe preserved in sedimentary rocks. Although Cuvier was prepared to accept that one such 'révolution' had been a flood about 6000 years ago, he was a man of the Enlightenment. Although a practising French protestant, he believed in keeping science and religion apart, and did not attribute supernatural causes to the 'révolutions'.

Diluvialism was flourishing in Oxford in the 1820s. The first professor at Oxford (appointed in 1823, twenty years before a Chair was established in Trinity College Dublin), William Buckland (1784-1856), was an avowed catastrophist whose aim was to return natural history 'to the explicit service of religious truth'. He was greatly influenced by Cuvier, but believed that the Flood had a supernatural cause. Buckland was the banner holder and flag waver of the diluvialists, and offered no less than nine points of proof of a global flood. He believed that rivers were incapable of producing their valleys and instead appealed to the awesome energy of Noah's flood. He noted loose sediments resting above bedrock in Scotland and believed them to have been formed by deposition from a global deluge. He also believed that the bones and teeth of animals in caves were dragged there by hyenas and had then been covered by a layer of mud washed in by the waters of the Flood. Buckland seems fairly close to religious orthodoxy to me, but he was criticized by certain clerics at the time for taking liberties with the Holy Writ. As the historian of science Gordon Herries Davies has suggested[9], Buckland has the dubious honour of being the last British geologist of note to attempt to reconcile geological discoveries with a literal reading of the Scriptures.

Yet Buckland, despite this bad press, did not have a closed mind, and was forever attentive to new observations. He became seriously concerned at the lack of human fossils in diluvial deposits. Buckland had effectively given up on supporting the Biblical account of Noah's Flood with geological evidence by the time he met the Swiss Louis Agassiz, who showed that the supposed diluvial deposits were for the most part glacial, having been left by extensive glaciers during previous ice ages. Buckland therefore abandoned his ideas of a single, cataclysmic, global flood, and so did his followers, the idea collapsing like a pack of cards. Buckland needs to be assessed in the light of the intellectual, religious and social context of England in the early 19th century. Judged in this context, he should not be dismissed as a belated biblical literalist[10]. Buckland's position was relatively flexible compared to the dogmatic insistence on the literal truth of Genesis that later developed as a coherent body of propaganda known as 'fundamentalism'. To the fundamentalists of the early nineteenth century, Buckland and Cuvier were nothing short of heretics!

Biblical catastrophism set back Huttonian views of the Earth, but not for long. Charles Lyell took on his mantle, publishing his highly influential *Principles of Geology* in 3 parts, the last appearing in 1833. Lyell did not believe that a global flood had shaped the Earth's landscape. The Hutton-Lyell school of thought is known as *uniformitarianism*, meaning that the long history of the Earth can be understood by analogy with the essentially gradual processes operating today. Geologists over the next two centuries would find convincing evidence of the vastness of geological time, but evidence would also begin to accumulate that there was indeed a beginning. Geologists would collect evidence of gradual, slow change explainable from a knowledge of modern environments and processes, but they would also accumulate convincing evidence of sudden and violent events the likes of which have not been observed by modern man. So we have both uniformitarianism and catastrophism woven into a geological fabric.

The contrast of the two characters, Leonardo de Vinci and William Buckland, is fascinating. They are separated by over three centuries in time. They bracket an era of strenuous efforts to reconcile observations on Nature with the 'truth' of Scripture. Yet the irony is that Leonardo was three centuries ahead of Buckland in his intuition. If genius is something to do with being able to think the impossible[11], then Leonardo has to go down as a towering figure in the millennium.

How is it, then, that many people still believe in a literal global flood? The movement known as 'fundamentalism' evolved in the early 19th century to counteract a 'heretical' departure from biblical literalism. Leading figures in the launch of this movement were Granville Penn and George Bugg. Bugg wrote in his two-volume work *Scriptural Geology* (1826-7):

> 'I allow...that Sacred writers may be silent about science and even ignorant of it...They were under divine and supernatural guidance, and therefore personal ignorance in the writer is no defect; and error is impossible.'

The geologists who introduced a 'dangerous mysticism' into the interpretation of the Scriptures were regarded as the enemies of Christianity. Since geology provided evidence that contradicted a literal reading of Genesis, then geology must be wrong. The fundamentalist George Fairholm in his *General View of the Geology of Scripture* (1833) argued that all geology since the mid-eighteenth century had been misled by *continental* (my italics) philosophers, especially by the 'madness' of Buffon[12], who had relegated the Flood to one of many similar episodes that had affected the Earth. By the mid-point of the century, however, fundamentalist views on the Flood were weakening.

The Flood was accepted as local, as the German Georg Kirchmaier[13] had first suggested as early as 1667, but appeared universal because it affected the entire region known at the time (the Hebrew word used, *eretz*, may refer to earth, ground, even fields, and does not imply the entire planet). This view was even adopted by the conservative Church of England in *The Bible Commentary* of 1871. You might be tempted to think that this is the end of the story, but you would be wrong. Although a literal interpretation of Noah's Flood as a truly global phenomenon is no longer part of Church orthodoxy, it is still very much a part of the thinking of the great 20th century resurgence of fundamentalism, especially in the USA. The book *The Genesis Flood: the Biblical Record and its Scientific Implications* by J.C. Whitcomb and H.M. Morris was published in 1961 and has not been out of print since!

Megafloods: Stranger than Fiction

The occurrence of a truly global flood is dismissed by the scientific community and is not part of Church orthodoxy. However, the question remains of whether a historical flood actually took place, the story of which was passed down the generations by oral tradition. To pose this question is not a latter day return of the attempted harmonizing of Biblical sources with scientific discoveries. Instead, we are legitimately probing into the basis for flood legends appearing in many different cultural sources. Noah's Flood may now be shed of its literalist interpretation, but this world-picture of a global deluge shaped the course of thinking about the natural world for centuries.

Towards the end of the nineteenth century the geologist Edoard Suess (whom we have previously encountered in the context of continental drift) argued that Noah's Flood was due to an underwater earthquake located in the Indian Ocean, which caused a tsunami. In deep water tsunami waves have a small amplitude (up to 1m) and very long wavelength (about 200 km), and if you were sitting on a boat in the open ocean, you would definitely not realize that a tsunami had passed under you. However, as they approach shallow water, tsunami waves peak up. Initially, water is withdrawn from the coastal zone, but then returns with a vengeance. Suess believed a tsunami generated in the Indian Ocean raced up the Persian Gulf, peaked up by the funnelling action of this water body, and hit the low-lying areas of Mesopotamia, the possible location of the Flood legend. This theory is difficult to accept, for a number of reasons: a tsunami could not have travelled the approximately 1000 km from the shallow entrance of the Gulf in the Straits of Hormuz, because friction would have damped the energy of the incoming tsunami wave; the timing does not make sense since the period of tsunami waves is less than half an hour, which doesn't square with the idea of a long warning period, and with a prolonged inundation; the inshore penetration of a tsunami wave is strictly limited, usually measured in fractions of a kilometre to several kilometres.

Another possibility is a gigantic river flood, but there would have been little warning and the onset would be sudden, since the rivers of Mesopotamia, the Tigris and Euphrates, are fed by the snowfields of Armenia and Anatolia. This doesn't quite sound like the flood of Noah, where there was time to build an ark.

If a tsunami or a river flood was not the cause, then we need to think of something more colossal, so enormous that it was etched deeply into the minds of its survivors and passed down through the generations. In doing this lateral thinking, we visit western USA, take a detour to the English Channel and peer into a shrunken Mediterranean Sea before returning to Mesopotamia.

In the eastern part of Washington State, USA is an area of basaltic bedrock representing old volcanic outpourings of lava. It has been heavily eroded into a strange landscape of channel networks and basins, deep potholes and countless dry waterfalls. In 1923, J. Harlen Bretz (1882-1981) suggested that these features were produced by a catastrophic outburst from a lake dammed by glacial boulders and ice. He named it the Spokane Flood (from the town of Spokane, Montana), and termed this strange landscape the 'Channeled Scabland'. The idea flew in the face of uniformitarianism, the dominant world-picture at the time, and Bretz felt the full force of criticism. His detractors demanded 'a return to sanity and uniformitarianism'[14]. He defiantly stated that 'the one man rebellion was still alive', and continued his fieldwork.

By 1930, Bretz believed he had the solution to the Spokane Flood. He believed that as the North American ice sheet retreated between 18,000 and 13,000 years ago, the ice dam holding back a large glacial lake collapsed, allowing the catastrophic emptying of a huge volume of water from the lake. This water poured out at depths of 40 to 80 m over the landscape, scouring it deeply. The discharge of water, at about 40 cubic kilometres per hour, was thousands of times greater than some of the largest floods ever recorded on the Mississippi and Amazon Rivers. More than another two decades passed before Bretz's hypothesis was vindicated. In 1953, now approaching 70 years old, Bretz discovered enormous ridges of sand and gravel that looked like oversized ripples, similar in geometry to those seen in the dried up bed of a river. A hydraulic analysis of these giant ripples was later carried out[15], confirming the ideas of Bretz. It was also shown that there had been multiple floods in the Channeled Scabland area, of which the Spokane Flood was merely the most recent. Periodic flooding results from the inherent instability of ice dams blocking the drainage of pro-glacial lakes such as Lake Missoula.

It took half a century for the hypothesis of catastrophic flooding exemplified by the Spokane Flood outburst to be accepted. Only the longevity of Bretz's life – he reached 99 years old – allowed him to see his views finally accepted. After receiving the Geological Society of America's highest award, the Penrose Medal, Bretz is alleged to have said to his son 'All my enemies are dead, so I have no one to gloat over'.

Lakes on the fringe of the North American ice sheet have now been shown to have repeatedly drained catastrophically and recent research[16] shows this to be a very widespread phenomenon throughout the northern hemisphere ice sheet margins during deglaciation. For example, glacial Lake Agassiz, which ponded up against the North American ice cap, spilled into the Atlantic Ocean (the last spillage taking place about 8400 years ago), and the vast Eurasian ice cap catastrophically drained through the ancestral Volga river system into an enlarged Caspian Sea.

When J. Harlen Bretz first discovered the fantastic tracery of incised valleys, dry waterfalls and giant sand and gravel ridges that spoke of an ancient catastrophic flood from dammed glacial lakes, it was not known how common these features were. In 2007, a geophysicist (Jenny Collier) and geologist (Sanjeev Gupta) from Imperial College London published a letter in *Nature* providing stunning support for an earlier proposal that the English Channel, a shallow tidal seaway separating southern England and northern France, was once the site of a large catastrophic flood that burst out from an ice-dammed lake situated in the southern North Sea[17]. They were able to image the shape of the bedrock surface beneath the choppy waters of the English Channel using a high-resolution sonar technique[18]. This detailed new map showed features eerily similar to those caused by the Spokane Flood. There were large flat-bottomed valleys cut into hard bedrock, with islands that were streamlined in shape, and deep longitudinal grooves looking like giant scratch marks. The map contained the unmistakable features of a catastrophic flood.

The English Channel flood took place at some time during the glaciation that affected the northwest European area over the last million years. The drainage of the glacial lake occupying the southern North Sea cut the topographic bridge that joined France and England, severing migration routes for fauna, including early humans. During low levels of the sea at times of glacial expansion, Europe's largest rivers would have drained through the English Channel to the Atlantic Ocean, making use of the breach in the barrier. At high levels of the sea, at times of ice retreat during interglacial periods, the flooded English Channel connected with the southern North Sea, forcing the mouths of northwest Europe's major rivers to move back to close to their present positions[19]. Britain thus became an island[20], causing a major bifurcation in its anthropological and cultural development. This is worth pondering when standing on the vertical white Chalk cliffs of the Kent coast, overlooking one of the world's busiest seaways, picking out the vague colour contrast in the distance that is the coast of France.

The volume of water discharged from the land by catastrophic floods is not large compared to the volume of the stuff in the oceans. However, the water is fresh and its arrival is sudden and repeated. Pulses of fresh water from glacial Lake Agassiz in North America are thought to have caused a weakening of the conveyor belt of Atlantic Ocean water circulation. It caused a period of cool and dry climate lasting about 1300 years about 12 thousand years ago, called the Younger Dryas. This rapidly tipped the northern hemisphere into a period of glacial advance, vividly illustrating the possibility and effects of very abrupt climate change. The effects of the draining of glacial Lake Agassiz were felt all over the world in the form of climate change, extinctions of animals, and the development of early agriculture.

Going back in time a little further to 6 million years ago, during the time known as the Messinian (after the current-swept Straits of Messina that separate Sicily from the mainland of Italy), the Mediterranean Sea became isolated from the Atlantic Ocean. Sea levels in the Mediterranean Sea fell by 2 kilometres, as the water evaporated into shrunken saline remnants. All around the Mediterranean, rivers charged down steep canyons, delivering water to thirsty saltpans and brine pools. For example, the River Nile cut down deeply into its bed, to levels 2.5 km below present-day sealevel close to the future position of Cairo[21]. About 5% of the total dissolved salt of the world's oceans was deposited in a fraction of a million years[22] – that's more than 1 million cubic kilometres of salt.

The Atlantic Ocean repeatedly spilled into the western Mediterranean, and when the connection to the Atlantic was re-established at about 5.3 million years ago, Atlantic Ocean water is thought to have poured into the Mediterranean as a gigantic waterfall and cataract. Mediterranean water subsequently freshened as sealevels rose again.

In their fascinating book *Noah's Flood* [23] (1998), geologists Bill Ryan and Walter Pitman piece together the case for a catastrophic flooding of the Black Sea by the Mediterranean. They suggest that the Black Sea was once a body of fresh water lying well below (about 100 m) the level of the world's oceans. It was fringed by river floodplains that were fertile and which therefore offered a favourable habitat for human occupation – a kind of Garden of Eden. This lake fringe, claim Ryan and Pitman, was abruptly flooded by an influx of salty water, which must have rushed through the Bosporus from the Mediterranean, carving a deep channel and depositing a fan of boulders at its northern terminus. From the isotopic dating of the new arrivals of marine shells in the formerly fresh lake, the date of the catastrophic flooding was about 7600 ago, or 5600 BCE (Before Common Era, equivalent to Before Christ, BC) – close enough in timing to be a candidate for the

Noah's Flood legend.

An expedition to the Black Sea in 1999-2000 funded by the US National Geographic Society[24] discovered timber house beams, rough stool tools and ceramic storage vessels under approximately 100 metres of Black Sea water. The discoveries were made by taking images using a remote-controlled mini-submarine and by sonar from a surface ship. An old beach was identified by the National Geographic team, which included shells that have subsequently been dated as 7500 years old by the radiocarbon technique. Together, this provides convincing evidence for the former human (Neolithic) occupation of the fringe of a shrunken Black Sea when the water level was considerably below the level of the world ocean.

Ryan and Pitman believe the catastrophic flooding would have caused a major dispersal of peoples from this cradle of civilization around the Black Sea. Some of the fleeing peoples settled in the Mesopotamian area as ancestors of the Sumerians. The story of the great flood would have been handed down from generation to generation by word of mouth. Such stories were the source for the Flood narratives, according to Ryan and Pitman. In order to facilitate the handing down of the stories as a means of maintaining their cultural identity, artistic devices would have been used to aid the memory. This may explain the elements of myth woven into the narrative.

As an example, the leading of the animals two by two into the ark, and rain lasting forty days and forty nights, may be such devices. Certainly, if you think perhaps of your own childhood, children are very good at remembering these aspects of the story. Sometimes it's the *only* aspects they remember! But it brings the story to life. It aids memorization and retelling. I do not believe that these phrases were ever meant to be taken literally. Indeed, it would have been quite impossible for Noah to take two of *every* living creature on board the ark. Did he make a quick trip to Australia to grab a few marsupials? Did he make sure he got two of every bacterium? Of course not. The point of the Biblical story is that God extended both his judgment and his plans for a new beginning to *all* of his creation, not just to Man.

Since Ryan and Pitman first proposed the idea of a catastrophic flooding of the Black Sea in their 1998 book, there have been a number of studies to check its accuracy, including a cross-disciplinary research project funded by UNESCO and the International Union of Geological Sciences[25] on the history of the Caspian-Black Sea-Mediterranean corridor over the last 30,000 years[26]. Instead of a major catastrophic flooding, the flow of water between the Aegean Sea and the Black Sea is thought to have been gradual or reversing and unremarkable[27]. No evidence could be found for the major rise in sealevel in the Black Sea suggested by Ryan and Pitman[28]. Their alluring story of a Noah's Flood is therefore disputed and controversial.

Shed of its literalist interpretation, Noah's Flood remains a powerful metaphor of a change in lifestyle through the ushering in of a new regime, as shown by the dove bearing an olive leaf. The notion of a global flood profoundly affected geological thought for hundreds of years, acting as a powerful world-picture through which a growing amount of natural science was initially refracted. Finding the theatre of the flood in the Black Sea region may prove to have been a false start, but the rooting of flood legends in actual historical events remains a strong likelihood.

A Plague on Both your Houses

In 2013, the Natural Environment Research Council, the government agency in the UK responsible for funding research in the Earth and environmental area, announced additional targeted funding in response to the increasing flood risk of the UK. Minor though this risk may seem in relation to the hazardous predicament of dwellers on some of the world's great river floodplains, infrastructural losses in the UK caused by flooding were rising.

Flood disasters have been happening for as long as Man has chosen to settle in large number in vulnerable places. Those most vulnerable to flooding live in river basins and at the coast, where threats arise from landfall of intense cyclones and from tsunamis generated by fault movements and volcanic eruptions under the ocean. Floods are recognized as the most potent natural hazards on Earth[29] – a greater risk even than earthquakes. Pakistan, Bangladesh and China regularly experience devastating flooding affecting millions of people each year and an estimated 2.5 million people died in the flood of the Huang River in China in 1887.

Ancient writings, including Greek mythology, tell us of frequent floods. Plato's Critias writes[30]:

'Many great deluges have taken place during the nine thousand years, for that is the number of years which have elapsed since the time of which I am speaking; from that time to this'—the soil which has kept breaking away from the high lands during these ages and these disasters, forms no pile of sediment worth mentioning, as in other regions, but keeps sliding away ceaselessly and disappearing in the deep'.

Sometimes, a flood disaster can be reconstructed from cryptic sources, as I believe is the case with an event that affected the River Nile and its delta region, as recorded in 'Exodus', the second book of the Judaeo-Christian Bible. Although some writers question the historical authenticity of the admittedly bizarre events recorded in 'Exodus'[31], the underlying context behind the narrative merits closer examination[32].

Exodus

The curious story of the plagues of Egypt[33] is portrayed as a fundamental turning point in the history of the Israelites, and records events taking place over three thousand years ago. In the Exodus narrative the Israelite leaders and spokespersons Moses and his brother Aaron went to the ruler of Egypt with a request on behalf of the large number of Israelites who had settled there as workers or slaves. The request to be allowed to go out into the desert and to offer sacrifices to their God was refused and resulted in further persecution, which caused the Israelites to become disheartened and restless. Perhaps it is not immediately clear why the Israelites would have wanted to get away from prying Egyptian eyes in carrying out sacrifice and worship, but it would be rather like killing a cow in front of a Hindu or eating a pig in a synagogue. Israelite sacrifices would be offensive to the Egyptians since sheep or rams were sacred to the Egyptian God Amon-Re.

The story continues that Moses and Aaron returned to Pharaoh to renew their request. There follows a battle of might, enacted symbolically through the participants' staffs, which became snakes, or to translate the Hebrew more liberally, a lizard or even a young crocodile, which would perhaps be appropriate for a location close to the banks of the River Nile. Pharaoh would still not agree to the request of the Israelites. And so at this impasse, and as the stakes are being ominously raised, we begin the account of the plagues.

There is a common pattern to the plagues, which can also be seen as 'blows', 'strokes' or 'signs' from the Hebrew. In each case, the Egyptian Pharaoh is asked to allow the Israelites to be able to go into the desert to offer sacrifices to their God. On his failure to agree, the plagues occurred. The pattern is that the plagues stop when Pharaoh appears to agree to the requests of the Israelites, but then changes his mind, leading to the following plague. The literary devices used are evident from the plagues being divided into three blocks of three, with the first of each block delivered to Pharaoh in the morning as he went to the Nile, as a warning of the consequences of not releasing the Israelites. First there are the plague of blood, the plague of frogs and the plague of gnats. Then there are the plague of flies, the plague on livestock and the plague of boils. Thirdly, there are the plague of hail, the plague of locusts and the plague of darkness. Finally, there is the death of the firstborn animals and firstborn sons in Egypt.

In the plague of blood, the river turns to blood, the fish die, the river stinks and its water becomes undrinkable. All water in streams, canals and even buckets, are turned to blood. Pharaoh was unimpressed.

Seven days passed before the plague of a swarm of frogs (or if used onomatopoeically, croaking animals including toads), which come out of the waters of the Nile and invade the land. This was followed by the plague of gnats that came from the ground, infesting man and animals. Gnats may include a variety of insects such as sand flies, fleas or mosquitoes.

In the second of the three blocks of plagues, the Pharaoh is then confronted with the prospect of a plague of flies that would afflict the Egyptians, but not the Israelites camped at Goshen, a fertile region in the eastern part of the Nile delta. The next threat was against the Egyptian livestock, which died in the fields. Since the Egyptians worshipped many animals as deities, this was a direct confrontation of the Egyptian religion.

The narrative tells of Moses taking a handful of soot from a furnace and tossing it in the air, causing a fine dust to settle over Egypt, leading to festering boils on man and animals. There may well be some symbolism in the taking of soot from a furnace, which may have been a kiln used for making bricks. In this sense, the furnace symbolized Israel's captivity and oppression in Egypt. The tossing of soot from the furnace into the air sounds like a defiant gesture of resistance.

Starting the third block of plagues, the next plague was in the form of a major hailstorm, so that anything left out in the fields would die. Thunder, rain and lightning accompanied the hailstorm, but it did not fall in Goshen where the Israelites were concentrated.

The Egyptians were warned of a plague of locusts that would devour what was left after the hailstorm. An east wind blew all day and night and in the morning the locusts arrived, turning the ground black and devouring everything. Then darkness fell on Egypt for three days. Since the Egyptians worshipped the sun god Ra, this was another direct insult to their deities.

Clearly, we have come to the end of the diplomatic road. In a flush of anger, Moses told Pharaoh of the last plague, the death of all firstborn males in Egypt, including cattle. This was the ultimate disaster. The blood of slaughtered lambs was to be placed on the doorframes of the Israelite houses where the lambs were to be eaten that evening. That same night the firstborn Egyptians were killed, but God passed over the houses with the blood of the lambs on the doorframes, which is why the Jewish festival celebrated today is called the Passover. Pharaoh summoned Moses and told him to leave with all his women, children and livestock. The Israelites included about 600,000 men on foot according to the text of 'Exodus'.

It is of course tempting to dismiss the entire set of plagues as outright mythology, conjured up as some kind of loose connection of calamities that had been experienced previously. Yet there is an order to the plagues that is coherent and sensible. Before embarking properly on further investigation, however, it is worth noting that some scholars believe that the narrative of the plagues originates from two separate accounts, each with 7 plagues, and that their amalgamation gave a grand total of ten plagues[34]. Secondly, it is important to realize that the account of the plagues was passed down through the generations orally, so has features to aid memory and visualization in the retelling – especially exaggeration and the use of repeated motifs.

We start with the Nile turning to blood, a phrase used in Sumerian and Egyptian mythology[35] as meaning water unfit to drink. Two explanations spring to mind, both requiring us to read this as 'turned *as red as* blood'. The first explanation is of a river flood caused by heavy rains in the catchment region, and the second is of a flood from the sea.

Catastrophic river floods are common in arid regions after sudden, violent rains. In dry regions with only sporadic rainfall, the bedrock crumbles to a loose, deep red coloured soil. The River Nile is fed by water from an area of about 3 million square kilometres, water flowing from mountainous headwaters in Ethiopia and Sudan to the flat desertlands of Egypt, before entering the sea, where it has built a triangular shaped, low-lying delta. From headwaters to delta is a vertical elevation difference of 3.8 km[36]. It is highly likely therefore that heavy rainstorms in the uplands of southern Egypt, Sudan and Ethiopia during late summer and early autumn caused very large amounts of sediment-choked water to be funneled into a relatively narrow channel, causing overtopping of the banks of the Nile. The river water would have turned into something approaching tomato soup.

Any degradation of the inevitable admixture of organic material, or algal blooms driven by high nutrient levels, would have removed oxygen from the river water and caused the river to stink. Since all the irrigation channels and canals were fed from the Nile, they too would have been contaminated, and water drawn from the river for domestic consumption and stored in buckets would have been undrinkable. Egyptians dug along the sides of the river for drinkable water, where the filtering action of seepage through permeable river bank sands and silts would have removed the suspended particulate material, rather like the filter paper in a coffee maker. The digging for fresh water at the sides of the river shows incidentally that the river could not have turned literally to blood, otherwise it couldn't possibly have been filtered clean.

Typical storm-related flood events are initially marked by a peak in the concentrations of sediment suspended in the flowing water, followed later by a peak in river discharge, leading to overtopping of the river banks. Choked by the high sediment concentrations and possibly low oxygen levels, aquatic animals would have fled from the water onto the dry land, causing the plague of frogs one week after the river water turned red. Out of their natural habitat, and perhaps infected by bacteria picked up from the Nile water, the frogs died in large numbers. The plague of frogs would have also had some symbolic value to the Egyptians, since the frog or toad was deified in the God Hapi and the goddess Heqt, who assisted women in childbirth. They were therefore a fertility symbol.

Imagine the breeding of gnats in the flooded fields of the Nile. As the waters receded, the floodplain would have been left sodden and veneered with red sediment and piles of dead and decaying frogs. Flies bred and multiplied. These flies were obviously of the biting variety. Carrying disease, these flying insects would easily infect livestock that had been brought back into the fields after the floodwaters had receded. Humans also seem to have been infected, individuals coming out in dark-coloured boils, reminiscent of Anthrax, which infects animals, particularly domesticated species such as sheep, goats and cattle, and can be transmitted to humans.

Thus far, the plagues fall into a pattern recognizable as a flood disaster followed by its aftermath. The plagues can also be explained if we envisage the events as being related to a volcanic eruption, such as the Santorini volcano in the Aegean Sea. The eruption caused a dense plume of volcanic ash to be ejected as much as 48 km into the atmosphere[37], so would have been distantly visible to the inhabitants of the Nile delta region.

The first major impact would have been a tsunami wave, which would have reached the Egyptian delta region in less than an hour after the eruption, causing extensive flooding of the low-lying coastal plain, contaminating drinking supplies, and killing freshwater fish in channels and lakes. It has been suggested that the red colouration was caused by blooms of tiny aquatic algae known as dinoflagellates. These 'red tides', which may last for months, may have been promoted by iron-rich dust blown in from the Santorini eruption, falling into the ocean like suddenly adding fertilizer to a garden lawn. A red tide driven onshore by tsunami waves would kill many fish, and cause frogs to try to escape contamination.

The acid, irritating airborne dust and ash would have arrived shortly after the tsunami, and may have been retold as gnats coming from the ground. Swarms of insects would have entered houses, animals in the fields would have choked and poisoned, and boils and blisters would have developed on the skin. The disturbance to the atmosphere by the volcanic eruption would have caused a hailstorm accompanied by heavy rain, washing down large quantities of dust and ash. Darkness may have resulted during the heaviest intensity of ash cloud over the delta region.

The linkage of 9 plagues in a logical sequence using the Santorini eruption is tidy, though not fully convincing, as we shall see. The other possibility is that a flood disaster is concatenated with other woes that happened afterwards. For example, the hailstorm may have taken place in January or February of the following year, because we are told that the hail fell at the time when the barley and flax were in flower, but the wheat and spelt (an inferior form of wheat which grew well in poorer and drier soil) had not yet ripened. So as much as 5 or 6 months may have passed since the flooding of the Nile. East winds blowing from the Red Sea, Sinai and the Arabian 'empty quarter' occur mostly during March and April. These winds may have blown in locusts with large appetites, as happens today in the drier parts of Africa. In the Exodus account, a change in direction of the wind, so that it now blew from the Sahara, dispersed the locusts, but it may have brought sandstorms from the giant sand seas to the west to cause the plague of darkness. The sheer concentration of dust in the atmosphere blots out the Sun's brightness and warmth.

Discriminating between these two 'natural' explanations is hampered by uncertainty of the timing of the sojourn of the Israelites in Egypt, and of the plagues in particular, compared with the timing of the Santorini eruption. We now turn to this question of timing.

The Santorini Eruption

Santorini is one of the best-documented volcanic eruptions known to Man[38], both geologically and archaeologically. It is probably best known for the Bronze Age eruption that buried the town of Akrotiri[39], though volcanic activity has been taking place at the site for hundreds of thousands of years. The legend of Atlantis, in which a whole city sank beneath the sea in a single day and night, may be based on the effects of the Santorini eruption on the first great European civilization, the Minoan, which dominated Crete and the Aegean world at this time[40]. The island of Crete is located one or two hundred kilometres to the south of Santorini. The Minoan civilization developed on Crete from about 3000 BCE and was distinguished by its large palaces with magnificent frescos, and elaborate pottery, metalworking and jewellery. The civilization declined from about 1700 BCE, probably as a consequence of a chain of events punctuated strongly by the effects of the Santorini eruption.

That the eruption of Santorini might have affected the inhabitants of Egypt was first questioned in 1971[41], and suggestions were made that the 10 plagues might have been a direct or indirect consequence of the Bronze Age eruption[42]. Studies of the volumes of volcanic products in the area surrounding Santorini led scientists to suggest that the bulk of the 60 cubic kiometres of erupted material was ejected into the atmosphere and transported far from the island. Santorini ash has been found in abundance in cores recovered from the seabed of the eastern Mediterranean and the Black Sea. A layer of Santorini ash has even been found in a small mountain lake in western Turkey. The Santorini ash was transported by winds that dispersed the material to the northeast, east and southeast of the volcano.

The age of the Minoan eruption has long been a subject of controversy, with estimates deriving from radiocarbon dating, from tree ring information and from ice core studies. Radiocarbon ages of plant remains in volcanic deposits associated with the Minoan eruption have yielded ages spread about an average of 3600 years, or *circa* 1600 BCE[43]. Radiometric dating of grain and timber samples from Akrotiri on the island of Santorini suggest a ^{14}C age in the range 1500-1650 BCE[44], but the uncertainties are fairly large, perhaps as much as a hundred years. A prominent peak in sulphuric acid in core from the Greenland ice cap has been dated as 1645 ± 20 BCE. This has been tentatively correlated with the Minoan eruption. Analysis of tree rings shows that bristlecone pines in California and oaks in Ireland both show severe frost damage in 1628 ± 2 BCE, which may relate to the climatic effects of the Santorini eruption.

The problem is in unambiguously attributing the tree ring and ice core data to the Minoan eruption of Santorini, rather than to other major eruptions at about the same time period (for example, Vesuvius, Italy, or Aniakchak, Alaska). The best available evidence therefore situates the Minoan eruption in the middle of the seventeenth century BCE. 1628 BCE is commonly taken as the date of the main eruption of Santorini.

The timing of the Israelite sojourn in Egypt is more problematical. Biblical sources combined with archaeological evidence give a range of possible dates for the Exodus[45]. The details need not concern us, but on balance, the most plausible solution is that the Exodus from Egypt took place 480 years before Solomon's reign over Israel (1 Kings 6:1), which places it at about 1447 BCE. If so, the Santorini eruption pre-dated the Exodus by a century and a half. The Santorini eruption may have had important historical impacts, but apparently is not implicated in the 10 plagues of Egypt. The linkage between the Santorini eruption and the Exodus illustrates what might be called the 'smoking gun syndrome'. The tendency is to link two events that are known to have occurred, which looks sensible in terms of cause and effect, despite a lack of precision in the dating of the two events.

If the main Minoan eruption of Santorini was not responsible for the events that led to the flight of the Israelites from Egypt, then can it be related to another historical event? There is another Biblical narrative (Genesis chapter 47), which tells the story of the Israelite Joseph. It is a story of famine, to which we will shortly turn.

Whatever the 'natural' explanation, the plagues signify a disaster or multiple disasters that would have had a prolonged and vivid impact on those involved. It was an 'annus horribilis'. The population of the Nile delta region would have been in a state of increasing panic, and the Egyptian rulers would have tried to alleviate the effects of the natural disaster, with or without prompting by Moses and Aaron. People would have fleed from the hotspot of the disaster. When at a safe distance, their collective memories would have been fused and the results passed down orally, as the Israelite people consolidated behind Moses, the man who had come to the fore during the natural disaster, and perhaps who now retrospectively is given a key role in the legend.

Famine

The story of the Israelite Joseph is associated with the establishment of an emerging nation of Israelite people in Egypt. He rose to prominence in Egypt as an interpreter of dreams. Joseph interpreted Pharoah's dreams as indicating that there would be seven good years of plentiful harvests followed by seven years of famine. He was appointed to take charge of the careful planning and stockpiling of corn during the years of abundance in preparation for the lean years ahead. He was joined in Egypt by his family, including his father Jacob.

As foretold by Joseph, Egypt and surrounding regions experienced seven years of plentiful harvests then plunged into famine. We are told that both 'Egypt and Canaan wasted away because of the famine'. We should not be too concerned with the periods of 7 years of fruitfulness and famine, since the Old Testament in general, and the book of Genesis in particular, is not particularly concerned with exact time periods. Instead, numbers are commonly used to indicate completeness, or perfection. Thus, the number seven in the case of the famines merely emphasizes the fullness of the years of plentiful harvests and the outright severity of the years of famine.

Famine may have many causes rather than one, but there is often a trigger for calamity. What was that trigger in the case of the Egyptian famines in Joseph's day? An obvious possibility is the failure of rains in the upstream drainage basin causing the Nile to fail to flood, leading to drought. The Egyptians relied on regular-as-clockwork flooding of the fields during the wet season (August). Alternatively, catastrophic floods of the Nile may have caused such massive inundation of the valley that crops were unable to be sown and harvested. However, it is possible that there was a completely different kind of trigger, unrelated to the Nile, since it is clear from the text in Genesis that a wide area was affected by the famine, implying that whatever the cause, its effects were felt in a region far beyond the Nile Valley. We must look for a mechanism that can have very widespread effects. Such mechanisms are most likely climatic or atmospheric in origin. So the question is, what could have perturbed atmosphere and climate sufficiently to cause the famine?

As we saw with the occurrence of plagues prior to the Exodus, we need a chronology or time frame in which to snugly fit Joseph[46]. There has been some difficulty in pinpointing Joseph's era accurately[47]. Using historical sources, the length of the Israelite sojourn in Egypt was approximately 215 years. On this basis we can say that Joseph was in Egypt from approximately the year 1682 BCE[48], and that the famines began shortly after. However, this is not certain, and Joseph's sojourn is dated differently according to the Egyptian dynasty in which it is placed. For example, the scanty evidence in Egyptian manuscripts of the disastrous famine in Egypt at the time of the Israelite Joseph points to it occurring in a period of particularly poor documentation left from the Hyksos 15th Dynasty of Egypt (1633-1525 BCE)[49]. On the other hand, David Rohl in his book *A Test of Time*[50] argues that Joseph's time in Egypt was during the long reign of Amenemhat III in the late 12th Dynasty, which according to his chronology ended in 1633 BCE.

In looking for a mechanism that could have perturbed atmosphere and climate sufficiently to cause the famine of Joseph's time, we have a prime candidate – the Minoan eruption of Santorini.

Volcanic eruptions are well known for their impact on climate. Volcanic eruptions are commonly viewed in terms of lava streams snaking down hillsides and bubbling cauldrons in craters. However, the damage and loss of life caused by lava flows is trivial compared to the effects of the airborne ash and volcanic gases. Volcanic ash thrown up into enormous clouds above the volcano can be dispersed in a number of ways. The ash column may collapse under gravity, producing a very fast moving, ground-hugging current full of incandescent dust. These currents are capable of engulfing entire villages, taking their inhabitants by surprise. The unlucky inhabitants of Pompeii discovered this to their cost in AD79 when Vesuvius erupted. Ash may also be ejected into the atmosphere and dispersed over large and small distances by winds. The heaviest ash falls are closest to the site of the eruption. However, the greatest dispersion of ash occurs when it is erupted high into the stratosphere, where it is caught up in the global system of jet stream winds, which carry it all over the world. In this way, volcanic eruptions can have severe effects on global climate.

Volcanic eruptions have been unequivocally linked to famine. The Haze Famine of 1784-1786 in Iceland is an example[51]. Large volumes of degassed SO_2 fell as sulphuric acid, killing vegetation and resulting in large losses to livestock (75%) and humans (24%). Dust from the eruption fell to the ground as far away as North Africa. Temperatures in the northern hemisphere are believed to have fallen by 2°C as a result of the eruption[52]. In the case of the Haze Famine, we have a largely agrarian community living at the edge. A similar vulnerability may have been found in Egypt during the time of Joseph.

We have tantalizing glimpses of the connection of the Santorini eruption, the decline of the Minoan civilization and famines in Egypt and the Near East. Severe climatic disturbances are therefore linked to one of the biggest volcanic eruptions in human history, coinciding with the time of Joseph in Egypt, but not of Moses nearly two centuries later.

The accounts of plagues and famines therefore have their origins in events that actually happened, and which can be linked to flood and volcanic disasters. In looking for natural explanations, we easily fall prey to the smoking gun syndrome, since there is a human tendency to link events that serve conveniently as cause and effect. On reflection, the Israelites were not so much different in trying to make sense of the events in which they were embroiled and through which their identity was defined. Except that their smoking gun was a saving act of God.

10 DEEP TROUBLE: THE WITCH'S CAULDRON

In which the deep circulation of the Earth is contested amid accusations of zombie science . . . .

The Power to Surprise

The Geological Society of London hosted a 2-day scientific meeting on 1-2 September 2011 with the title 'Dynamic Topography'[1]. London was basking in an oblique, hazy warmth as summer gave way to autumn, lawns lay full of dew till late morning, and maples in Hyde Park were curling and colouring in resignation as the biological clocked turned. Outside in Piccadilly was the bright and perfectly normal chaos of the centre of an international megacity. Cars, taxis and buses, plus a few unfortunate cyclists, were backed-up from Piccadilly Circus to the darkened red brickwork, Portland stone dressings and elegant lines of St James's church, designed by the architect who changed London's skyline forever, Sir Christopher Wren. Traffic commonly draws to a halt as far as the ornate shop front of Fortnum & Mason, with its spectacular clock. If delegates of the conference had the urge to include a lunchtime recital by a chamber choir or string quartet, they could do worse than walk just 100 metres to St James's, or alternatively, if they felt the need to empty their pockets, they could do it very efficiently in half an hour at Fortnum & Mason by simply crossing the road. The Geological Society is therefore well situated.

Inside, amid the quiet classical elegance of Burlington House, a topic as astonishing as anything the geological sciences can throw up was being discussed – the idea that deep, very deep, beneath our feet the Earth is slowly circulating like a witch's cauldron, and that the elevation of the surface of the Earth records in some way that deep circulation.

When you consider that it was not long ago when one would be incarcerated for denying that the Earth is flat, it will come as a surprise to those who have since come to believe that the Earth is the shape of a sphere that they too are mistaken. The Earth bulges out at the equator as it spins on its axis, but much more interestingly, the surface of the Earth goes up and down like a beating heart. These undulations are nothing to do with colliding plates pushing up mountain ranges. The undulations are seen even in the floor of the ocean. What's more they may fluctuate in their elevation, width and position over time, so are ever changing. Some believe that they fluctuate with a regular pulse. This seems incredible, but seldom does anything that appears at the time to be common sense turn out to be right. This goes for the present as well as times past.

At the end of the Middle Ages the universe was conceived as a bounded system with the Earth at its centre. Nicolaus Copernicus (1473-1543) suggested that the Sun rather than the Earth was at the centre of the solar system. This upset the clerics, though it is difficult to understand why, since nowhere in religious scriptures are these astronomical niceties discussed. Galilei Galileo (1564-1642) invented a refracting telescope and found strong evidence supporting the Copernican idea of the rotation of the Earth around the Sun. He so annoyed the Roman Catholic Church that he paid for his views by his inquisitorial trial in 1633 and house arrest for the last 8 years of his life. At his inquisitorial trial he was forced to recant his view that the Earth moved around the Sun. It is said that he then famously muttered under his breath 'and yet it does move'. Kepler (1571-1630) moved the debate still further from the geocentric view when he found on the basis of telescope work that the orbits of the planets were ellipses rather than perfectly circular.

It's remarkable now to think that these discoveries sparked such controversies. How could the organized Church possibly object to the orbits of the planets being ellipses rather than circles? Perhaps because an infinitely powerful God wouldn't create anything as imperfect as an ellipse when a nice circle would do? Yet the Bible is absolutely silent on whether the Earth orbits the Sun or not, and it says not a word on the geometric shapes of the orbits of the planets. Copernicus was a believing Catholic, yet Church leaders found his views heretical. Conventional wisdom was in the hands of the Church.

Isaac Newton (1642-1727) applied laws of gravity and inertia to the movement of the planets, showing that the planets continue in their inertial motion around the Sun held by the force of gravity. He was accused by the famous mathematician Gottfried Leibnitz of being 'subversive'. Apparently, the idea that planets could orbit without some external push from the hand of God undermined religious belief.

That the age of the Earth might be measured in billions of years also came as a big surprise. And how absurd is the idea that the continental foundations on which we build our cities are splitting and joining and in perpetual motion? We have more recent surprises. It takes a lot of imagination to believe in black holes and dark matter and curved space-time, and even more in the small worlds of particle physics.

We are used to surprises, especially in the geological sciences, whose practitioners have been, as Tennyson put it in *Locksley Hall*, 'nourishing a youth sublime, with fairy tales of science, and the long result of Time'. Just as we were getting used to the idea that the deep circulation of the mantle is manifested in the ever-changing topography of the Earth's surface, dissenting voices are articulating an alternative world-picture.

The Witch's Cauldron

The Earth is made of a number of concentric shells that result from its differentiation after it was initially formed from a collapsing ball of cosmic dust. The light elements floated to the top, giving the continental crust with an average thickness of 40 km and with the average composition of granite. The heaviest elements, such as nickel and iron, gravitated to the centre, where temperatures and pressures are high enough to produce a liquid outer core at greater than 2800 km depth. In between is the vast bulk of the mantle. The outer skin of the Earth comprising the crust makes up just 0.018 of a percentage point of the total volume of the Earth, and even less

(0.009%) when measured in terms of mass. In contrast, the mantle makes up 83.6% of the Earth by volume, and 67.7% by mass. The crust is like the skin on an apple when compared with the size of the Earth.

The continents lie dispersed in oceans floored by basalts, formed at submarine mountain ranges known as mid-ocean ridges. Every year, another 2.8 square kilometers of ocean crust is formed in this way. The formation of new oceanic crust at these ridges causes the oceans to expand and to push the continental masses around like reluctant passengers. If continued over time, this process alone would result in an ever-increasing ocean area, but the process is counterbalanced by the disappearance of ocean crust by descending beneath other oceanic or continental plates in a process known as subduction. Due to this balancing act, the area of oceanic crust has probably not fluctuated strongly over time and has remained close to 61% of the total surface area of the Earth, it's present day value.

The Earth is therefore made of a discrete number of plates, with different proportions of oceanic and continental crust, which are in constant motion relative to each other. With modern technology, the velocities of the world's plates are known accurately[2]. We know that the rate at which India is colliding head-on with Asia is about 25 mm per year, that coastal California is sliding past the rest of North America along the San Andreas fault system at nearly twice this rate, and that the Atlantic Ocean is getting wider by about 45 mm each year on each side of the mid-ocean ridge. These rates are similar to the rate at which our fingernails grow. It may not seem that dramatic, but it means that oceans can open and expand very quickly in geological time, growing to 100 km in width in little over 10 million years, and continental fragments can speed across the surface of the Earth covering a distance of 1000 km in a trifling matter of 40 million years. These are the rates of plate tectonics. No wonder earthquakes and volcanoes occur where plates collide and subduct.

The outer shell of the Earth is like the congealed skin on porridge or custard that is allowed to cool. It is sufficiently rigid that it moves as a coherent slab or 'plate', detaching over the fluid-like rock underneath. The oceanic crust continues to cool as its gets older and is pushed further away from the mid-ocean ridge, and eventually sinks like a capsizing ship at a subduction zone. As the oceanic plates sink into hotter surroundings, gravity acts on their higher density and drags them down into the abyss, causing them to slowly plunge through the mantle. The slabs either suffer the fate of being recycled, thereby contributing to the uneven texture of the mantle, or they continue to sink until eventually reaching a graveyard around the outside of the core.

Plate tectonics is a theory that has become a paradigm supported by an immense amount of observational data. But why should we believe that the Earth beneath the plates is being stirred like a witch's cauldron? There are probably two main reasons.

First, by studying very small variations in the speed of waves generated by earthquakes, which traverse the deep Earth, it is possible to make three-dimensional pictures of the distribution of 'fast' and 'slow' regions of the Earth's interior[3]. These tomographic images are similar to the scans used in medical diagnosis. The fast and slow regions in the deep Earth either represent density differences due to small changes in the composition of rocks, or density differences due to variations in temperature. These small but important density differences affect the speed of seismic waves. If the density differences reflect temperature variations, then these are best explained as due to a deep, large-scale circulation in the mantle. If the density differences reflect changes in rock composition, it would imply no such flow, but instead a deep interior that was not uniformly mixed.

Adopting the first of these alternatives for the moment, since this is conventional wisdom, a three-dimensional map can be made that contains information on circulation patterns enacted over long periods of time. So slow is the circulation that we see in one picture an accumulated history of countless millions of years of flow, like watching an entire movie in one image.

Second, experiments in the laboratory and on powerful computers suggest that the Earth has a temperature distribution that makes it unstable and prone to a large-scale circulation known as convection. You see convection when boiling jam in a saucepan that is heated from below and cooled from above. Less well known are the convection of water in a lake, heated from above, and the convection in the atmosphere that causes tropical downpours in late afternoon after a hot day. Convection is triggered when the temperature gradient is too great for viscous forces that resist movement to withstand. Convection experiments and computer models of the Earth show upward-moving hot masses of rock and downward-moving cold masses that have been linked to the three-dimensional pictures derived from seismic wave speeds.

The conventional view is that the high temperature gradient between the hot outer core and the relatively cold slab graveyard causes convective instabilities to form, which are thought to grow into hot conduits of rock that ascend towards the surface. Reaching the base of the plates that comprise Earth's outer shell, they are thought to spread out like a subterranean mushroom cloud, arching up the plate, heating it, fracturing it, and allowing huge volumes of the mantle to melt and flood to the surface as volcanic eruptions. These are the 'plumes' of the planet, which with youthful vigour replace to the surface the volume lost by the mournful descent of cold slabs to the interior of the Earth. Together, they make up the slow stirring of the witch's cauldron.

However, although regarded as conventional wisdom, this model of Earth's deep circulation has been challenged, and the role of plumes in particular has been questioned[4]. It is claimed that hotspots, where volcanoes punch through the crust, are not tethered to the deep mantle via underlying umbilical cords[5]. Instead, they are better thought of as 'melting anomalies' since their unambiguous signature is the solidification of melts below the surface and the eruption of lavas above it. Caltech geophysicist Don Anderson believes that these melting anomalies are fundamentally a result of the varying 'fertility', or tendency to melt, of mantle rocks, rather than a result of temperature variations in a circulating but evenly mixed mantle.

The conference at the Geological Society in 2011 was aimed at providing a further angle on Earth's deep circulation in the form of its surface topography. If the surface topography caused by processes in the mantle could somehow be measured, it would be possible to compare the results with the predictions based on the three-dimensional maps constructed from earthquake wave speeds. The race is on to better understand the topography caused by deep circulation, since to do so would be one of the most significant advances since the formulation of the plate tectonics paradigm.

Blobs and Buoyancy

I have often heard it asked, what actually is 'Dynamic Topography'? My short reply, while bathing in conventional wisdom, would run something like: 'it's the vertical displacement of the Earth's surface generated in response to flow deep within the Earth'[6]. These words seldom placate the enquirer, and I am starting to realize why. For after two days of pleasant incarceration in Burlington House in 2011, I discovered that the experts were not exactly singing from the same hymn sheet. Does it matter how we define 'dynamc topography'? What's in a name anyway? Surely a rose by any other name would smell as sweet?

The conference room was not full of Capulets and Montagues in tribal conflict but of a veritable 'who's who' of the world of geodynamics. The lack of a clear definition of the meaning of the term 'dynamic topography' was the dark and silent secret of the meeting.

So here is the crux of the matter, and you will see that it is more than how you like 'tomato' and I like 'tomahto', otherwise we would call the whole thing off[7]. When gravity acts on density differences it produces a force known as buoyancy. Gravity acting on a hot air balloon relative to a cool outside atmosphere causes a positively buoyant force that makes the baloon rise. The size of the buoyancy force depends on the density difference between the hot air inside the balloon and the cool air outside, the volume of air in the balloon and the gravitational acceleration of the Earth. The bigger the volume and the density difference, the bigger the buoyancy force. A sense of the buoyancy force can also be appreciated by holding a ball under the water of a swimming pool. The bigger the ball and the lighter the ball in terms of the air inside, the harder it is to hold the ball under water. You are counterbalancing the buoyancy force.

Now imagine a source of buoyancy deep in the Earth caused by the presence of a less dense lower layer beneath a denser upper layer. The lower layer wants to bulge upwards caused by its positive buoyancy. Viscous stresses transmit this upward acting force, which is opposed and balanced by gravity acting on the upward deflected surface. Note that buoyancy causes a deflection of previously horizontal layers. If you like to think of it this way, the positive buoyancy of the 'blob' acting as the source of the density contrast is counterbalanced by the negative buoyancy of the upward deflected bulge. So this is what dynamic topography is. The source of the density difference promoting buoyancy may ultimately be caused by flow, and the blob of buoyancy may well be moving, causing a dynamic response by the resistance of the surrounding viscous material, but dynamic topography is not really dynamic at all. It is calculated at an instant in time as simply a static balance between the buoyancies of the blob and the deflected surface that it generates.

If we wanted to calculate the dynamic topography from a known internal distribution of buoyancy, we would need to know the distance of the source of buoyancy (the blob) from the Earth's surface and the viscosity of the intervening material. This is why if we could accurately measure dynamic topography, it would help in the assessment of the viscosity structure of Earth's mantle, a kind of Holy Grail for geodynamicists. Knowing Earth's viscosity structure would enable geodynamicists to solve a range of currently intractable problems to do with how the Earth works.

However, it's not quite that simple. As every student of geology knows, the Earth's topography is due to more than buoyancy contrasts in its deep interior. In fact, Earth's topography is principally a reflection of density variations within the outer rigid plates that are in relative motion with each other. The topography of high mountains, for example, is possible because of the presence of deep roots made of relatively light rocks, a situation caused by thickening during the collision of tectonic plates. The Himalayas have snow on them not because of dynamic topography, but because they are underlain by 70 km of light, granitic rock, part of which belongs to Asia and part to India. To identify the component of topography caused by deep buoyancy, it is necessary to remove the more important component originating within the plates, but the procedure of obtaining this residual topography is more problematical than it sounds.

'Dynamic Topography' turns out to be elusive to measure, problematic to model, and perhaps even ambiguous to define. Essentially, the topography referred to is not truly dynamic, not in the sense of, for example, the surface elevation changes associated with the great currents of the ocean such as the Gulf Stream[8]. A sausage lying forlornly on a plate is not dynamic simply because it was manufactured on a machine with moving parts. It only becomes dynamic if it is flying through the air and hits you in the eye, thankfully a rare occurrence outside of school canteens.

However we define dynamic topography, this field of enquiry brings benefits whether you are looking downwards or upwards. Looking downwards, it has the potential to better constrain the Earth's distribution of viscosity. In addition, the record of dynamic topography over geological time periods provides a 'narrative' that allows the tempo of change in the mantle to be better appreciated. Looking upwards, the interaction of plate movements and mantle flow gathers the continents into supercontinental assemblies, heats them up and causes them to disperse - a fundamental feature of the planet we live on.

At a less grand scale, but of immense interest, dynamic topography may explain the ups and downs of continental interiors. Uplift generated by the deep circulation of the mantle may cause rivers to react by etching new drainage networks or reinvigorating old ones[9]. Dynamic topography caused the reversal of drainage direction of the mighty Amazon River – formerly, it drained to the Pacific Ocean, guided by dynamic topography related to the eastward subduction of the Pacific plate beneath South America. On the other hand, subsidence generated by deep circulation causes some continental areas to sag downwards[10]. Deep circulation is therefore probably the main factor in controlling the flooding of continents from which the history of global sealevel is calibrated[11]. In the ocean, dynamic topography may cause gateways and barriers to open and close, affecting ocean circulation and driving climate change[12]. The amount of dynamic topography may fluctuate as the strength of the underlying plume pulses[13]. Such pulsing may cause variations in the supply of sand to the deep sea derived by erosion over the topographic bulge[14]. The melting associated with hot regions in the mantle immediately beneath the plates is the main candidate responsible for massive outpourings of volcanic rocks that have been implicated in the occurrence of mass extinctions[15]. The slowly circulating witch's cauldron deep beneath our feet therefore plays an important role in a wide range of Earth processes today, just as it has in the past.

Testing the Plume Hypothesis

The existence of hot, narrow plumes arising from the outer boundary of the core has been used as an explanation of provinces of high volcanic activity known as hotspots, including linear chains of volcanic seamounts in the ocean, and topographic swells on land and in the sea bed. However, the plume hypothesis has been criticized and cited as an example of zombie science. It is argued that the plume hypothesis fails several predictive tests. If this is the case, should the plume hypothesis be abandoned, or is there room for a modified version?

The presence of plumes of hot material originating deep in the Earth was first proposed in 1971 by W. Jason Morgan (1935-) of Princeton University[16]. He made use of the work of Canadian geologist and geophysicist John Tuzo Wilson, who had proposed that the upwelling limbs of convection systems in the mantle were responsible for volcanic hotspots on the Earth's surface. In this model, as the plate migrated above a plume fixed in position in the mantle and rooted at the core-mantle boundary, it left a trail of volcanoes of increasing age. At one end of the hotspot track should be a currently active hotspot, whereas the other end of the track should be located where the rising newly born plume first impacted on the overlying plate. Commonly cited examples are the hotspot track joining the volcanic island of Réunion in the Indian Ocean with the Deccan region of western India, or the linear chain of Emperor-Hawaii seamounts in the central Pacific Ocean.

Mantle plumes are also invoked to explain the occurrence of volcanoes far from the margins of plates where plate tectonic theory suggests volcanoes should occur. Instead, hotspots occur far from plate margins in both the continents and oceans and their distribution seems to bear no relation to plate boundaries. The plume hypothesis has also been used to explain the formation of Large Igneous Provinces (LIPs)[17], containing colossal volumes of rocks derived by melting of the mantle. Examples are the 250 million year old Siberian LIP of Russia, with a volume of 4 million cubic kilometres of basalts, and the 65 million year old Deccan LIP of India, which has up to 1 million cubic kilometres of basalts. These sites of rapid and voluminous volcanic outpourings cannot be explained easily using the conventional plate tectonic model.

Laboratory experiments predict that rising plumes, comprising a wide head fed by a narrow, long conduit, should spread out into mushroom shapes of about 2000 km diameter beneath the overlying plate. On reaching shallow depths, the hot mantle material is believed to melt by a reduction of pressure (decompression), causing the surface eruption of flood basalts. The impingement of mantle plumes on the base of overlying plates has also been invoked for the break-up of supercontinents[18]. If the plume hypothesis were shown to be incorrect, it would topple a view that has been fundamental to the understanding of the dynamics of the Earth for over 40 years.

Each of the main lines of evidence in support of the plume hypothesis has been countered by opponents and explained in terms of plate tectonics. The details need not concern us, but it is worth noting that two world-pictures explain the same 'evidence'.

There are therefore two end-member positions, the plume hypothesis and the plate hypothesis, and little indication of a merger of aspects of the two opposing paradigms. Need the debate be so polarized? Scientists by nature take up adversarial positions, while diplomats do the reverse. Indeed there is a cachet in behaving with self-confidence and hubris in science.

Not all believe that plates and plumes are mutually exclusive and envisage them both operating in the dynamical system of Earth's mantle[19]. That dynamical system involves two boundaries where there are rapid gradients of temperature, termed 'thermal boundary layers'. The first is the hot boundary layer around the core, which causes narrow rising columns to rise to the surface of the Earth as plumes. To compensate for the upward movment of mass in plumes, broad, slow, cold downwellings are implied to exist in between the rising plumes. The second is the cold boundary layer near the surface of the Earth, which constitutes the plates. These slabs of rock form at mid-ocean ridges, cool, subduct and descend into the mantle where they are reheated. There is therefore a second style of convection related to the plates. The fact that plume locations are not, in general, related to plate boundaries (Iceland being an important exception) suggests that the two convection styles are more or less independent.

The circulation related to plates is the dominant style of convection. The controversy is not therefore about plumes overturning the plate tectonics paradigm. It is more about the importance given to plumes, from important yet secondary, to zero.

Those opposing the plume hypothesis, who are in a minority challenging 'conventional wisdom', have more to play for. The web site www.mantleplumes.org managed by Durham University geophysicist Gillian Foulger, covers a very wide range of topics, and commendably publishes different viewpoints. As of 5 November 2013, 653 scientists had contributed to the site. Gillian Foulger has won an award (Price Medal, Royal Astronomical Society, 2005), partly in recognition for her work on this site, and has also published a book on *Plates versus Plumes* [20]. The existence of this web site has undoubtedly stimulated a vigorous debate that has brought together specialists in geophysics, geodynamics, geochemistry and petrology.

Future years of research will see a resolution of the controversy. Will the plume hypothesis be finally put to rest after a period of zombie existence? Will the plume hypothesis supporters reassert their position? Will a merger of the two camps take place? The answers to these questions are currently unknown.

The controversy of plumes *versus* plates continues to rage. At its most polarized, it is the fight by a growing minority against a conventional wisdom that has outgrown its usefulness and exists in a world between life and death. At a more nuanced level, it is about holders of two opposing world-pictures each of whom can legitimately appeal to a range of observations and modeling studies that appear to support their case. The controversy energizes the debate and focuses minds on what would constitute a clinching test from the points of view of both camps. This can only be a good thing. Yet the adversarial science of constructing opposing hypotheses runs the risk of missing the possibility of a middle ground – that it is a case of 'plates *and* plumes' rather than 'plates *versus* plumes'. But nobody particularly likes a compromise. It sounds like diplomacy, not science.

11 SALAMI SLICING EARTH'S OUTER SHELL

In which a model is accepted, without sight of evidence, based on the high status of a major corporation . . . .

Strewn by the Million

On the chilly eastern flank of England is a town, a small town. It is situated where the coastline turns its back on its Scandinavian and Teutonic cousins, particularly when the jet stream arcs southwards at the impulse of the mere flapping of a butterfly's slender wings in some distant hemisphere. Cold Arctic air rolling off the greyness of the North Sea frequently compels the inhabitants of this small town to turn up coat collars and shiver. It is also situated where the River Alde attempts to enter the sea, but is deflected listlessly for miles along the coast where, when it has almost given up hope, an opening presents itself and fresh and salt are sweetly mingled.

Standing on the well-sorted brown shingle, the beach stretches like an arrow northwards to the eroding cliffs that turn someone's loss to someone else's profit, from depletion to accretion. Strong, rusted cables connect the waterline with weather-beaten wooden shacks close to a small concrete wall at the back of the beach, where small fishing boats safely perch away from the devouring tide that sweeps this coastline with metronomic regularity. Fresh herring, crabs and Dover sole are advertised on the fronts of the shacks, though few customers stop amid their bracing and strictly linear walks along the backshore.

The main shopping street of Aldeburgh ('old fort') is set back and exactly parallel to the seafront, protected from the easterly winds by a block of stoic but slightly seedy hotels and various brave but uninteresting dwellings. Window glass has that murky, translucent appearance caused by the accumulation of North Sea salt. In more clement climes, as at the coastal resort at the end of the tramline in Adelaide, South Australia, the waterfront would be lined with adventurous architectural experiments in glass, bold rooflines and quirky balconies, but the conservative English psyche prefers mock Tudor to Frank Lloyd Wright. Nevertheless, our Victorian and pseudo-Victorian buildings have the very beneficial effect of allowing the townsfolk and visitors to this small English town to walk along the main street carrying out their business in some degree of comfort. No doubt Benjamin Britten, Aldeburgh's favourite son and composer resident, did so in his time. It is to him that a stainless steel Scallop several metres high, standing on the brown shingle beach, is dedicated. It seems to have exercised paint daubing vandals many times over, though I doubt that their activities are a protest at the discordance of Britten's operatic pieces.

Walking in linear fashion north of Aldeburgh we go in search of the source that nourishes the beaches of the town with brown flints. The flints are generally hard, nobbly and rounded, with brown or bleached exteriors, but inside they are glassy black, with arc-like fractures. There are few things in this world tougher than a flint pebble. Now I appreciate that the search for the origin of the brown flint pebbles on Aldeburgh beach does not quite have the romance of discovering the source of the Blue Nile, nor does it have the imagination of J.R.R. Tolkien's *Lord of the Rings*, but it's pretty interesting nonetheless, for countless English men with handkerchiefs as head coverings tied by four small knots at the corners have sat on the brown shingle with their wives and children and not had the slightest idea of what they were parking their rears on, and what they were slinging distractedly into the waves, and what was swallowing up the small coins falling from their pockets with such ease.

In our search for the source of the brown flints, we are obliged to stop for a moment at the lost village of Dunwich, now beneath the waves, having kept a safe distance from Sizewell B nuclear power station on our route northwards. Dunwich also gives its name to a heath, with a thick, matted undergrowth of heather and gorse that camouflages small families of deer, and through which oaks, birches and the occasional pine poke to break the flat skyline. A walk down to the deserted beach reveals the heath to be underlain by sands, almost as friable as the day they were formed at the edge of a shallow sea not unlike the present-day North Sea, but 2 millions of years ago, just before ice began to periodically engulf the British Isles. The cliffs of ancient sands are wearing back by the attack of the North Sea, so that the elevated heath is shrinking, and it will not be long before the white coastguard cottages atop the cliff suffer the same fate as the old fort of Aldeburgh and the sad village of Dunwich.

It seems inconceivable that Dunwich once boasted a population of about 4000, who built 8 churches, three chapels and two hospitals, and exported from the harbour the wool and grain of ancient Suffolk. All that is left is a row of brightly painted beach huts, a red brick pub appropriately called the Ship Inn, and a small terrace of assorted houses. There is a melancholy about this loss, as if some bizarre calamity had struck the luckless inhabitants of this eastern extremity of England. But the reality is more mundane, though equally chilling, for the village succumbed, in stages like a Greek tragedy, to the free rein of nature. To the sea[1], the tempestuous sea.

Most of the destruction of Dunwich took place in the 13th century, and the last remaining portion disappeared at the same time as the First World War loomed, enacted its ugliness and finished amid relief and grieving. The coast was a mile to the east during the village's heyday, so a simple back-of-the-envelope calculation suggests that the sea advances by the length of a cricket pitch every ten years. For those readers unfamiliar with the strange and quintessentially English sport of cricket, the 22 yards of a cricket pitch is not much different to the distance of 18.39 metres between a pitcher and a batter in baseball. Off the Norfolk coast, a small distance further north along our linear walk, coastal look-outs built of ugly concrete lie askew in the shifting sands, pounded by surf, scoured by currents, testament to the retreat of cliffs since a second 20th century world war. The coast is far from the static frontier of an island nation, to be defended with Churchillian vigour. Frontiers are a human invention, and a disquieting illusion.

From these Norfolk cliffs come the brown flints that nourish the strand at Aldeburgh, first scraped by ice streams, then dropped, left like garbage, strewn carelessly, falling in landslips as the sea undercuts and devours, transports to new temporary resting places, and to a final oblivion who knows where. The indestructible pebbles are graded in size from north to south along their zig-zag route dictated by wind and tide of a Nature intent on orderliness. They are Earth's great survivors, left piled high on stormy beaches, rattling in contact together as backwash flows over and through them, defying time's arrow.

Strewn by the Million

What hard stone is this,
Strewn by the million beneath my feet,
Dark glass, brown weathered rind,
Piled up where wave and river meet?

From where comes this stone,
Where was it first crystallized,
Micron by micron from leaching brine,
Now drifted in zig-zags on reversing tide?

For how long has this palm-sized stone,
Held silent stories of universal time,
In colour, shape and rounded form,
That Nature briefly intersected hers with mine?

How big a world is in this stone,
Trapped incognito within blackened glass,
Stretching infinitely beyond my vision,
Yet into waves so carelessly cast?

The 2 million year-old sands in the cliffs and underlying the heath are elevated many metres above present day sealevel. Sealevel has varied enormously over the aeons of geological time, far more than the ten or so metres seen at Dunwich. There have been times in the past when continents have been inundated over thousands of kilometres by seas over 100 m higher than those of today. Seas have also been much lower, causing the margins of continents to be carved into great incisions deeper than the

Grand Canyon.

These submarine grand canyons exist around the edge of the
Mediterranean Sea, carved out long before the journeys of Aeneas and
Odysseus, but still relatively recently in geological terms. Just over 5 million
years ago the Mediterranean shrank to a salty remnant of its present state,
with sea level some 2 kilometres below the present. Rivers rushed in
torrents towards the bleak salty interior, incising canyons like gigantic
wounds in the newly emergent landscape.

By looking back through time, sedimentary geologists are therefore
able to recognize the clear signs of sealevel change, whether small in scale
and often repeated as at Dunwich, or large in scale causing extensive
inundations or desiccations. However, to read this book of sedimentary
geology, it is necessary to be able to link sedimentary rock types of the same
age across broad regions. This might seem obvious today, but three
hundred years ago there was no geological map to consult, and the ages of
rock units were largely unknown, so little was known about how the rocks
exposed at different places at the surface of the Earth related to each other.
Three centuries ago, that was all about to change.

One Man and his Map

On a corner in the road, at the entrance to the small village of Churchill in
the English Cotswold Hills, is a small monument that seems out of place in
its sleepy, pastoral surroundings. It is made from a rough-hewn, locally
quarried limestone, which succumbs to the gentle Oxfordshire climate by a
modest growth of encrustations of lichen and moss. Turning the corner,
one sees an impressive Gothic church set within a neat low stone wall, with
its steeples reaching skywards, in perfect architectural proportions to itself,
but oversized relative to the settlement that it serves, as if history had
abandoned it, consigned it as a picturesque relic. If this church looks a little
familiar to the inhabitants of middle England, it may be because it is
effectively a carbon(ate) copy of the chapel at Magdalen College, Oxford,
just 20 miles distant. There must be many stories to be told about this small
village, not least as to how it ended up with a copy of Magdalen College
chapel. Only one of these stories is that an inconspicuous child was born
here of a blacksmith father in 1769. His name was William Smith; he is

known as the 'Father of English Geology' and he drew the 'Map that Changed the World'[2].

In the course of his work as a surveyor, planning the construction of viaducts, canals and railway bridges that can be seen today, William Smith noticed that the sedimentary layers that had accumulated over long periods of geological time were in an ordered vertical sequence and could be traced across the English countryside. In the days of William Smith, it was important to be able to know the relative positions of particular rock units, largely based on the distinctive fossils that they contained, and to be able to correlate them from place to place. Times may have changed, and new techniques have emerged, but this aspect of correlation is still a key role for the science of stratigraphy. We need to divide Earth's long sedimentary record into chunks that make sense in terms of time, at as sufficiently high a resolution that modern techniques will allow.

William Smith published the first geological map of Britain in 1815[3], though he drew a rough sketch of it over a decade previously. The original hangs behind a protective curtain at the base of a curling staircase in Burlington House, Piccadilly, the home of the Geological Society of London. A full-size copy adorns the walls of the café and meeting room on the third floor of the Department of Earth Science and Engineering at Imperial College – known to many staunchly as the Royal School of Mines. Ignored by most students preoccupied with their iPhones and tablets, and staff with their cappuccini, these magnificent prints propel the observer back in time 200 years.

William Smith's maps were plagiarized and sold cheaply, which caused him to go into debt. Having no family wealth to fall back on, and with a humble education behind him that prevented his easy circulation in learned society, he had no wealthy patrons. Consequently, he was imprisoned. His house and belongings were seized. Recognition came late to him, and I imagine that it was a small consolation for the hardships that he and his family had endured. He was awarded the Geological Society's highest medal in 1831, when he was referred to as the 'Father of English Geology'.

The William Smith monument was erected in 1891, well after his death in 1839. His white marble bust, which I pass regularly when using the staircase in the Royal School of Mines at Imperial College[4], suggests a proud man with a hint of sternness inclining to defiance. I suspect that his thick eyebrows may have narrowed and his forehead furrowed a little at the antics of new-fangled stratigraphers. He made his cross-sections, established his correlations and drew his geological maps on the basis of direct observation of what was there. He would not have been impressed with red noise[5], shredded signals[6], statistical models, fluxes and budgets that obsess the more quantitative stratigraphers today. But I believe he would have had no problem with the view of stratigraphy emerging in the 20th century that sedimentary rocks are not a continuous record of the shifting environments at the Earth's surface over time, but instead are in large part stratigraphic snapshots, or 'frozen accidents'[7] enclosed in the swirling mists of unrecorded time. In their own way, the gaps in sedimentary rocks are observable by an absence of record during which time passed without any sediments left as clues. These absences of record can also be correlated from place to place as well as can rock formations. A systematic recognition of the importance of gaps was fundamental to the new way of thinking about the sedimentary record in the 20th century, to which we now turn.

Birth of a New Paradigm

Far from being stationary, geologists recognized that throughout the vastness of geological time sealevel had gone up and down continuously. Present-day sealevel is viewed at an instant in time in a wayward oscillation made of different frequencies and amplitudes. This wayward oscillation, however, looks a little less wayward if we consider the last couple of million years. Over this time period, we know that sealevel has fluctuated strongly with a periodicity of about 100 thousand years, which can be linked to the changing extent of land-based ice. For now, it is enough to know that since sealevel is constantly moving up and down, the coastline with its distinctive sediments must also move landward and seaward. This leads to a characteristic architecture of the sediment types that are preserved as they are buried beneath new layers. That same architecture can be seen in successions of sedimentary rocks that are brought back to the surface to be seen in coastal cliffs, mountain sides and quarry faces, and can be imaged by sending seismic waves deep beneath the Earth's surface. The seismic waves are reflected back off the boundaries between different sedimentary units and returned to recording instruments at the surface. This resulting salami slicing of Earth's sedimentary record shows that it is far from the imagined parallel layers as you would see in a sponge cake or a pile of pancakes. Instead the architecture is rich in 3-dimensional detail and challenging in complexity.

It is essential to know as much as possible about this subterranean architecture in order to successfully explore for hydrocarbons. As a result, petroleum companies have invested heavily in building models of the architecture of the sedimentary rocks beneath our feet. The most influential of these models was developed in the mid-1960s, 1970s and 1980s by Exxon, the largest oil company in the world. It opened up a new field in the Earth sciences called 'sequence stratigraphy', since the subterranean architecture shows the sedimentary rocks to be organized into chunks known as 'sequences', rather like recognizing storeys in a building.

The field of 'sequence stratigraphy' evolved as a new paradigm and drew its power from the recognition of sequences from seismic investigation of underground geology. These sequences were correlated worldwide and assumed to be generated by the rise and fall of the sea, as at Dunwich Heath. The paradigm has attracted admirers and detractors in equal number. It is viewed by some as a corporate model, developed in secrecy and released without transparency. It is self-referential, it is claimed, and has developed its own language. One of the original protagonists of the hypothesis, Peter Vail, defended the idea retrospectively in 1992[8], having then left employment in Exxon, with the now-familiar sounding refrain:

'The business of research is new ideas. Few things, however, are more unpopular with researchers than truly new ideas'.

Are these the warning signs of a pathological science?

The Building Blocks of Underground Space

Regions of the Earth's surface are slowly, yet profoundly subsiding over long periods of time. Sediment accumulates in these subsiding regions, or 'sedimentary basins'[9]. They sink for a variety of reasons. Perhaps the most common reason it that the rigid surface shell of the Earth (the plate) is being stretched like a block of plasticene or Blu Tack™. If the stretching is great enough, the plate may break, allowing a new ocean to form. Stretching of the plate causes it to be heated, so after the stretching has stopped, the plate subsides over a period of about 100 million years as it cools back to its original temperature, causing slow and decelerating subsidence. The largest sedimentary basins on Earth are due to stretching, including the continental margins along the edges of the Atlantic Ocean.

Other sedimentary basins are formed by the bending of the plate, just as a swimmer bends a diving board or a plastic ruler is flexed absent-mindedly by a schoolchild. These basins due to bending are particularly well developed next to mountain belts caused by the collision of continents. They accumulate the sediments derived by erosion of the uplifting mountains, as in the river plains of the Indus, Ganges and Brahmaputra in Pakistan, India and Bangladesh.

Sedimentary basins are therefore the great repositories of sediment, and have been a striking feature of the surface of the Earth since plate tectonics began billions of years ago. The new sequence stratigraphy targeted the fine detail of the sedimentary architecture of these basins.

The original Exxon model was revolutionary when it appeared in a memoir of the American Association of Petroleum Geologists in 1977[10]. The essential idea is fairly simple. Exxon geologists assumed that the margins of sedimentary basins were subsiding, which over time makes room for sediments to accumulate. Sediments were assumed to be supplied to the basin margin at a constant rate. They further assumed that sealevel rose and fell periodically. Sealevel fall causes the coastline to advance into the ocean, shifting seawards a mosaic of depositional environments. Sealevel rise causes the reverse – the flooding of the land surface and the landward shifting of the depositional mosaic. A cross-section or profile of sediments deposited during one cycle of sealevel change resembled a sigmoidal-shaped slug, with a 'sequence boundary' formed by erosion during sealevel fall, and a surface representing the greatest extent of flooding of the land due to sealevel rise. Now a further crucial assumption was made, that the timing of the rises and falls of sealevel was the same over the entire world. Consequently, the record of sealevel rises and falls could be used as a global pacemaker or chronology - a potentially immensely powerful tool that might re-energize the understanding of Earth history.

This 'supertheory', weaving together several related hypotheses, has proven difficult to test. The New Zealander marine geologist Robert Carter[11] believed that the hypothesis could be looked at in terms of two linked but separable concepts: first, the nature of the architectural patterns observed in the sedimentary fillings of basins, and second, the integration of the timing of the observed events into a global chronology over geological time. Few would dispute that there is value in the observational method proposed by Exxon, though there would be profound disagreement as to the causes and non-uniqueness of those observed patterns. Regarding the second, the protagonists were clear and outspoken in their vision[12]:

'One of the greatest potential applications of the global cycle chart is its use as an instrument of geochronology'.

Peter Vail continues, futuristically,

'Using global cycles . . . an international system of geochronology can be developed on a rational basis . . [to provide] a more accurate and meaningful standard for Phanerozoic [last 543 million years] time.'

There can be no doubt therefore as to the missionary zeal of the main protagonists in the 1970s and 1980s. There is no sense of their hiding their light under a bushel, of making a short but incremental, tentative step of progress. The Exxon group saw this as a paradigm shift that would provide a new pacemaker of geological time.

Even a relatively trivial description of the key points that cause disagreement would submerge us in a morass of technicalities and even worse jargon, but that is not the purpose of this chapter. Instead, we are fundamentally interested in the world-picture held by the corporate originators of the sequence stratigraphy model. The growth and popularization of the Exxon paradigm was largely due to a particular blend of social factors[13]. Before we proceed to investigate the basis of this view, we need to return to the pre-history of the model.

Despite its jolting impact, the Exxon model did not spring out of a vacuum. What does? To say that a person who has made a discovery 'stands on the shoulders of giants'[14] is a truism, albeit a colourful metaphor. In the early years of the 20th century, the American geologist Joseph Barrell (1869-1919) carried out a thought experiment of a subsiding basin accumulating sediment with a superimposed sealevel that went up and down like a sine curve[15]. The sealevel variation might, for example, be due to changing volumes of land-based ice sheets, as the Earth has experienced 17 or more times over the last nearly 2 million years (at a recurrence time of about 100 thousand years). Depending on the balance between subsidence and sealevel change, sediment is either preserved or eroded. Consequently, although successions of sedimentary rocks, when seen in a cliff or quarry wall, look like a continuous record of sediment deposition, when viewed in terms of time, the record is full of gaps. The American stratigrapher Peter Sadler[16] put it this way in 1981:

'Continuity and steadiness of sedimentation are troublesome notions.'

The American stratigrapher Larry Sloss (1913-1996) recognized the major building blocks of stratigraphy across the North American continent, publishing a landmark paper in 1963[17]. He joined the faculty of Northwestern University in 1947, staying until his retirement in 1981. This association with Northwestern is worth noting in passing. He noticed that these building blocks were separated by long gaps near the centre of the continent, but that the gaps reduced when traced to the continental margin, where sedimentary rocks are more continuously recorded.

That the stratigraphic record is like the life of a soldier, involving long periods of boredom punctuated by short periods of terror, has been used, most flamboyantly by Derek Ager[18], to support the idea that stratigraphy is 'more gap than record'. If this is the case, our epic poem of the Earth is a book with many missing pages. Charles Darwin[19] recognized that it was thus so when he described the geological record as:

'a history of the world imperfectly kept, and written in a changing dialect; of this history we possess the last volume alone ... only here and there a short chapter has been preserved; of each page, only here and there a few lines'.

The idea of 'more gap than record' is today embraced by the concept of completeness, and was formalized in the early writings (1981) of Peter Sadler. Sadler proposed that the apparent continuity of sedimentation is an illusion based on the resolution at which the sedimentary record is sampled, and that the completeness of sedimentary successions differed according to the types of environments, such as alluvial plains, the coast and shelf, and the deep sea, in which the sedimentary layers were deposited.

In the few years before his death at the age of 83, Sloss emphasized that he did not believe global sealevel change was responsible for the major gaps in successions of sedimentary rocks deposited in continental interiors and margins, preferring to envisage patterns of uplift and subsidence driven by tectonic processes, including those originating deep below the surface in the mantle. He clearly had a sense of humour, for one of his last papers, published in 1991, is entitled 'The tectonic factor in sealevel change: A *countervailing* [sic] view', in which he referred to those who claim an ascendancy for global sealevel change as neo-Neptunists. Perhaps too much can be made of this, but I suspect that he made these remarks with a certain amount of reflection that the neo-Neptunist *extraordinaire* was Peter Vail, one of his former PhD students at Northwestern.

Peter Vail (1930-) was primarily responsible for the chart of global sealevel variation over geological time, and developed his early ideas from the sedimentary sequences of the interior of the USA, where he must have been strongly influenced by his professor Larry Sloss. He got a Masters and doctoral degree from Northwestern University, Illinois, in 1956 and immediately entered the oil industry, moving to Exxon Production Research Company (called Esso at the time) in Houston in 1965. He stayed there until retirement in 1986, when he joined the faculty of Rice University in Houston.

Instead of tectonic processes affecting the height of the continental surface, Vail believed that the long-term cycles (of order 100 million years in duration) in the elevation of sea level were caused by changes in the volumes of the underwater mountain chains known as mid-ocean ridges. When the Earth erupts new ocean crust rapidly at mid-ocean ridges, so that the ocean floor spreads away from the ridge quickly, the mid-ocean ridge increases in its volume, displacing ocean water over the edges of the continents[20]. It seems peculiar now that those invoking this process did not consider that there was also a loss of ocean water where the ocean crust and its overlying soggy sediment were subducted beneath the continents, but the balance between the two effects is still up for grabs. At the other extreme, Vail thought that the shorter cycles were caused by astronomical forcing, that is, to changes in solar radiation caused by the elliptical path of the Earth around the Sun and by changes in the tilt of Earth's axis of rotation. There is good evidence that this controls the periodicity of ice advances and retreats in the last couple of million years. Vail also recognized cyclicity at the scale of a few millions of years, for which the astronomical forcing mechanism is less certain. He also found such cycles at times in the geological past when most people believe the Earth had no land-based ice, which makes any external variation in solar radiation less effective.

The breakthrough by Exxon was to recognize (or more correctly to hypothesize) the existence of widely distributed gaps in the record of sedimentary rocks beneath the surface of the Earth using seismic techniques. The new 'seismic stratigraphy' was developed from a number of distant continental margins and continental interiors, revealing, in the view of the protagonists, an apparent common timing of stratigraphic events around the world. These events were therefore believed to be caused by global rises and falls of sealevel. This then is the potential added value of the Exxon model – it provides a methodology for recognizing architectural patterns in the underground geology of sedimentary basins, and critically, attributes changes in these architectural patterns to sealevel change acting globally. Consequently, the Exxon model turned from being a means of collecting observations to becoming a predictive tool for the chronology of Earth history. Depending on your outlook, this is either a change from a kindly Dr Jekyll to a sinister Mr Hyde[21], or an ugly duckling becoming a swan.

The Science of Corporations

The new sequence stratigraphy paradigm was developed in Exxon's research laboratories in Houston. It became accepted by a large part of the stratigraphy community outside Exxon[22]. Although the very influential publications of the Exxon group were not subjected to independent critical review, the work gave the appearance of being legitimized by the excellent graphics, prestige of the publishers, and reputation of Exxon. As a result, many geologists in academia and industry accepted the model and began using it without question in their work. Strangely, this in turn legitimated the sequence stratigraphy model within Exxon, who had gone luke-warm following the initial publication of results. By accepting the new paradigm, the external community, comprising geologists, geophysicists and engineers who worked for or consulted for other companies, and academics whose funding streams relied on oil company grants, had much to gain. The use of a new jargon, the sense of belonging to a new, exciting club, and the ease with which the paradigm could be made to work in application after application, contributed to an uncritical adherence.

The conventional distinction between academia and industry as two 'social worlds'[23] was therefore blurred. The enthusiasm extended to both the new methodology and the global chronology. The new observational methods of recognizing sequences, based on seismic imagery of buried sedimentary rock formations, and then on the equivalent formations exposed at the surface of the Earth, were adopted with alacrity. In addition, the global curve of sealevel variation was seen as the key to more successful exploration of many of the world's sedimentary basins in which oil and gas are found. The sociologist-geologist team of Charlene and Andrew Miall wrote in 2002 (page 308) that:

> '. . assumptions about Exxon's resources, scientific expertise, authority, and credibility served to override, for both the academic and professional geology communities, and concerns about the lack of data, replication and independent proof for the model itself.'

In the workplace, two principles guided the Exxon research[24]. First, a working environment was fostered that directed group efforts to finding solutions of predefined problems, and to agreement on the methods to be used and the types of phenomena to be studied. The tool was the new database of seismic reflection profiles upon which sequences could be recognized that were widely correlatable, thereby allowing a new time framework to be developed. Second, an environment was created where individual responsibility was allocated to those best able to tackle the problem. These individuals took responsibility in areas that overlapped with those of others, ensuring good communication. However, this also ensured that individuals had a strong vested interest in the success of the broader venture. The result was that the Exxon group was 'socially constructed'[25]. In socially constructed research, commitments are made to colleagues, funding organizations or employers that maintain belief in the enterprise even in the face of uncertainties, which are therefore buried out of view.

The success of the new paradigm was legitimated by claims that the new methods worked in practice. However, the basic observations were not released into the public domain, so outsiders had very little opportunity to contribute to the testing of the paradigm and thereby to share in its development. Instead, the development of the paradigm was carried out by insiders who had a role in its original construction. In this way, the global cycle chart of sealevel change was augmented by further data, but these data were chosen by insiders who had vested interests in the success of the model. The basis for such data selection was never made public. It may have been done with the intention of strengthening or consolidating the original breakthrough, but the likelihood is that the selection process was affected by 'human' or 'social' factors. These factors guaranteed that the data that were selected to provide a revised, improved version of the global cycle chart of sealevel were those that agreed with the notions of the Exxon scientists involved.

Ups and Downs of the Sealevel Paradigm

Despite these obvious warning signs, a generation of stratigraphers has used the Exxon global cycle chart as the time framework for their studies. Is this as risky as it first seems? Why, if it is risky, did so many geologists apply this time framework themselves? The answer probably lies in its effortless use, rather than entailing lots of hard toil; in the fact that many geoscientists were funded by hydrocarbon companies and took the route of least resistance to ensure continuing or future funding; and in the belief that the new global cycle chart offered a once-for-all solution that enabled exploration geoscientists to come up with an answer instead of agonizing over uncertainty. The paradigm continued sublimely, aided by the formation of a cluster of believers who built solidarity, kept up motivation and cited each other in published literature.

It is perhaps surprising that the Exxon paradigm has survived as long as it has without suffering some holes below the water line. A simple consideration of the driving forces for the deposition of sediment in subsiding basins demonstrates that the sediment supply from the land must be viewed as a variable, not a constant over time. In addition, subsidence caused by tectonic factors such as the bending, faulting or cooling of a plate, is spatially and temporally variable, and very complex where faults grow and interact with each other. These factors might be regarded as a bit of noise of nuisance value only, and in some cases with justification. However, in other cases, such complexities would require a different interpretation of the positions of high and low sea level. This in turn would invalidate some parts of the global cycle chart. In the years of increasing disquiet about the Exxon paradigm, this is precisely what happened; seismic stratigraphers such as John Underhill of Edinburgh University investigated key areas used to construct a part of the global chart of sealevel (such as the Moray Firth, Scotland) and claimed that many boundaries between sequences were due not to change in sealevel but due to fault movements[26].

More problems lay ahead. It was not appreciated by Exxon staff at the time of the development of the paradigm or supertheory that deep circulation in the mantle causes temperature variations, which in turn cause density anomalies that affect the elevation of the surface of the Earth. One of the most powerful ways of stimulating such circulation is by the subduction of cold slabs of oceanic plate beneath continental masses. Such cold slabs descend *en route* for the inner reaches of the Earth, but near the surface part of their descent to oblivion, they cause the surface of the Earth to subside due to thermal effects, perhaps by a few hundred metres at most. In the case of the Farallon slab[27], an old bit of the Pacific Ocean that descended beneath North America about a hundred million years ago, the slab travelled right under the continent before descending steeply under the eastern seaboard of USA. By doing this, the east coast of USA, which has provided key information for a large part of the global cycle curve, was evidently flooded by the waters of the Atlantic Ocean not because of global sealevel rise but because of subsidence due to the descent of the Farallon slab. To make matters worse, the west coast of Africa, which has also furnished key parts of the global cycle chart, did not experience the subsidence due to the Farallon slab, since the slab descends under eastern North America. In other words, there is no constant datum to which we can measure sealevel variations. To appeal to global sealevel variation as the dominant control on the timing of the boundaries between sequences, and on the architecture of the resulting sedimentary deposits, therefore requires an appeal to a control that is unknown, or at the very least uncertain.

There are now two schools of proponents peddling competing paradigms, with little contact between the two groups, viewing the same data but seeing different things. They are operating with different world-pictures.

These competing world pictures are particularly divergent when considering the Earth's time framework. Without a time framework, we risk making false connections between events and missing others that might be correlated. In our everyday lives, we depend on everyone using the same chronology, which is why many of us wear wristwatches synchronized to some standard.

The establishment of a time framework for the Earth's history during which fossils are abundant (the Phanerozoic) continues to be a goal for those wishing to analyze and make use of this narrative record, and as a basis for the exploitation of Earth resources. Geological consultancy businesses[28], for example, use a model based on the recognition of 125 sequences within the last ~600 million years of Earth history, which works out at an average of one sequence per 4 million years. Their form of salami slicing is based on extensive analysis of published literature, not in-house databases, so they cannot be accused of a selective use of proprietary data. The age assignments are based on the chronostratigraphic information of each stratigraphic surface of interest, which must be globally (across several continents) recognized, despite, they claim, the complicating effects of variations of sediment supply and tectonic subsidence and uplift. While being independently constructed to the Exxon curve of global sealevel, and despite the filters applied to extract true global variations of sealevel in the past, there will be those who question whether the boundaries between sequences can be treated as synchronous across the globe, and whether the resolution of the dating exercise is sufficient to rule in or rule out correlation[29]. With so many events to choose from, correlation with a global standard is guaranteed, or as Andrew Miall put it, there is 'an event for every occasion'[30].

The competing paradigms therefore rumble on, with major consequences for the efficient exploitation of the vast wealth of hydrocarbons and minerals beneath the surface of the Earth. The development of the sequence stratigraphy paradigm can be seen as a corporate exercise that gained fulsome acceptance by virtue not of the basic data, but because of the status of the originators. That many of the academic community were prepared to passively accept interpretations not independently verified shows the intellectual power that can be wielded by organizations situated outside of conventional academic protocols.

Salami slicing Earth's outer shell reveals the ravages of time. One is transported through the ever-changing landscapes of the Earth's surface, passing through millions of years of natural history like a high-speed train through suburban stations. Standing on the long beach beside the eroding cliffs of sand below Dunwich Heath, under a grey, lowering sky, and with the rhythmic beat of shingle in motion, I felt vividly a personal onward march of time. I felt the freakish and ephemeral intersection in this vast expanse of time of two hearts and minds and bodies. We hugged each other silently, she facing the sea and I the cliffs, as a small tear formed, hovered on the brink and descended, un-noticed as it quickly dissipated through the wet sand. Moments later a foaming rush of swash engulfed our temporarily etched footprints. I blinked, crouched, played with the last remaining rivulets as they meandered seawards, and picked up a small, corkscrewed twig of heather ripped from the heath. I rose, fumbled with my coat buttons, grasped her hand in one of mine and held the corkscrewed twig in the other as we slowly climbed the gently sloping beach and departed.

12 PUBLIC ENEMY NUMBER 1: CARBON CRISIS?

In which an unlistening public is sceptical of a scientific consensus, and environmentalists dream of a low-carbon economy . . . .

London Fogs

Reading in quick succession Victor Hugo's *Les Miserables* followed by almost any novel by Charles Dickens is not advisable. It leads to an abject despondency about the plight of the urban poor. Cities such as Paris and London have had their fair share of social problems before, during and after the Industrial Revolution, but there are few icons of environmental pollution as vivid as the smogs of London during the nineteenth and twentieth centuries.

London's infamous fog of December 1952, also called the 'Big Smoke'[1], was due to airborne pollutants, mostly released from the burning of coal, which combined with the windless conditions of an anticyclone to encase the capital city in a yellow-black shroud for four days. At the time, there was that phlegmatic acceptance that the English are, or used to be, well known for. But when the environmental reports were written later, it was believed that perhaps as many as 12,000 had died, mostly from respiratory problems. The inhabitants of London were well used to urban smogs. The New York Times wrote in 1871 that the population was 'periodically submerged in a fog the consistency of pea soup'. Charles Dickens (1812-1870) famously wrote in 1852 in the epic of social criticism

Bleak House (chapter 1) about 'Fog everywhere . . . Fog up the river, . . . Fog down the river . . Fog on the Essex marshes . . . Fog on the Kentish heights . . . '. Not that London has a monopoly of pollution of urban air. Driving southwards from the exaggerated beauty of Zion National Park, high on the dazzling Colorado Plateau, one becomes aware of a yellow-grey smudge above the silhouetted hills that looks like grime on the front windscreen. The grime doesn't wipe off, and soon the smog of Las Vegas becomes obvious, appearing all the more dramatic after experiencing the crystal sky of the plateau. Whether Los Angeles or Beijing, Las Vegas or Shanghai, the visible expression of Man's pollution of the air is urban smog.

The London smogs are iconic, but they were not global. The world is currently in the grip of another carbon crisis, but this time the pollutant is insidiously invisible. I am referring to the burning of fossil fuels to release the gas carbon dioxide, which almost all scientists believe drives global climate change through the greenhouse effect. This contemporary and pressing problem, involving the polluters, the consumer, international panels of experts, environmental advocates, regulatory authorities and Governments representing the public, is most starkly portrayed with the conflict imagery of the *Carbon War*[2], and less confrontationally a decade later as the carbon 'Challenge'[3]. Are the holders of conflicting views of the climate change debate arranged like soldiers lining up in some medieval battleground? Why do different people come to different conclusions about essentially the same data? What are their world-pictures?

The Cycling of Carbon

The Earth is teeming with life, from ocean floor to mountain peak. Life has a habit of getting everywhere, including the most inhospitable of places like the darkest ocean, the highest mountain and the driest desert. In the oceans alone, 26 billion metric tons (a metric ton is 1000 kilograms, and a billion metric tons is a gigatonne) of organic carbon is produced annually. This is distributed over 361 million square kilometres of seabed, making an impressive 72 tons per square kilometre. The preponderance of life on our planet implies a global circulation of carbon, the key ingredient in the tissues of living things.

Plants and animals build tissue and skeletons using carbon, most of which returns to the ocean or atmosphere when the organism dies. But in certain situations, part or all of the carbon is not oxidized, leading to a long-term accumulation of carbon, as in swamps, wet soils, lake bottoms and especially the sediment of seabeds. In the ocean, 0.4% of the total organic production is preserved in marine sediment, which doesn't sound very much, but such is the vigour of life, amounts to 100 million tons of organic carbon each year. The rest disintegrates, dissolves or oxidizes and returns into the global circulation of carbon.

Burial of carbon in these settings takes carbon out of circulation. There were times in Earth history when the burial flux of carbon taken out of circulation was particularly high, for example, at the time of the coal forests of the Carboniferous period (about 320 million years ago). During burial, the carbon is heated and compressed. If heated moderately, the organic-rich sediment releases liquid and gas hydrocarbons, which migrate upwards towards the Earth's surface where it is seen in the form of seeps. Most of it is lost to the atmosphere in this way, but some is trapped *en route* to provide the accumulations of carbon treasure that remain sealed below the surface for millions of years. At higher temperatures, the carbon turns to coal. If we take buried oil, gas and coal and burn it for energy, carbon is released quickly into the environment and joins the reservoir of carbon in circulation. The Earth's atmosphere and oceans are receiving an overdose of carbon. It is on a carbon binge. The key question is whether this matters. The Earth has experienced high levels of atmospheric carbon dioxide concentrations in the past and seems to have got over it. Will it do so again, or are we in big trouble?

The world inventory of buried coal, oil and gas, that is, the amount formed over all of geological time amounts to some 10,000 gigatonnes of carbon (not counting shale gas). The size of this inventory is staggeringly large when you consider that burning of these fossil fuels accounts for less than 10 gigatonnes per year. So far, it is estimated that human activities have released between 300 and 500 gigatonnes of carbon, which also seems small compared to the total amount that is potentially available. However, this is not the point. The rate of burning of fossil fuels is a much more significant amount compared to the reservoir of carbon in the atmosphere in the form of carbon dioxide (only 700 gigatonnes). The air we breathe has carbon dioxide concentrations measured in parts per million. Consequently, our relatively feeble use of fossil fuels (compared to what is left to exploit) has a relatively large effect on atmospheric composition because the atmospheric reservoir of the gas is so small. There is therefore enormous potential for mankind to increasingly pollute the atmosphere, and thereby the oceans, from the burning of fossil fuels. We will not run short of fossil carbon for as long as it matters. If ever there were a time for restraint - a rarely used characteristic of the human race, this is it.

Carbon dioxide is also absorbed by the waters of the world's oceans[4]. Currently, about one third of anthropogenic emissions of carbon (2 gigatonnes) are dissolved into the oceans each year. It enters the surface oceans and is pumped into cold, deeper water. This takes the villainous gas 'out of sight and out of mind', but I'm afraid there is a price to pay and a loan to be paid off. The price to pay is that the ocean waters become more acidic by the dissolution of all that carbon dioxide, just like the effect of dissolved carbon dioxide in soft drinks. Think of that when you are next enjoying your glass of agua frizzante. The debt to pay off is that the oceans can't keep on acting as a giant reservoir of carbonated water. Carbon dioxide uptake will eventually refuse to cooperate any more – probably at about 5 gigatonnes of carbon per year by 2100[5]. After that point, emissions into the atmosphere will have no escape valve. We will have both turned the oceans acid and be set on a path to a dangerous acceleration of global warming.

Put bluntly, human activities are game changing when it comes to the carbon cycle, as repeatedly recognized by the Intergovernmental Panel on Climate Change.

Changing Climates

I doubt that there has ever in the history of scientific endeavour been such a diversion of funding into one issue of societal concern than the targeting of climate and climate change in the last two or three decades. The only rival is the directed science effort at times of war. Governments, especially in technologically advanced countries, who provide much of the money and infrastructure for basic research, have ring-fenced climate change for special funding while at the same time trying to safeguard other innovative basic research. Scientists working in areas that are outside of this ring fence have felt the wintry chill of having to find other ways to support their research. They are ingenious enough to succeed, but the concentration of resources on climate change serves to emphasize what a very serious problem it is perceived to be.

Yet, all do not agree. There are many sceptics. They range from those engaged in outright denial to those who accept that human activities are implicated in climate change but doubt the apocalyptic predictions of some of the opposing camp. Environmentalists have been highly effective in warning about anthropogenic climate change and in their advocacy of a greater emphasis on renewable sources in the energy mix. However, the oil industry is not quite the *bête noir* that environmentalists portray. Even if it were, the sinners can become saints by directing their colossal expertise towards the removal of carbon dioxide from power stations and its safe storage underground. In short, the oil industry can hardly be blamed for providing society with what it wants (I am careful to avoid the word 'needs'). Consequently, it is a rather second rate sinner, and it should not be treated as a saint for capture and underground storage of carbon dioxide, since if it does this, you can be sure that it will be at some profit to its shareholders. It is time to wise up, forget the conflict imagery, and work towards international agreements on the basis of the very best understanding of this human experiment on the planet's climate system as possible.

If we want to do something about anthropogenic climate change, we must firstly understand how the planet's climate system functions. In broad terms, the geological record, the narrative of what has happened before, provides a conceptual or intellectual framework for this understanding. The geological narrative underscores that we do indeed have a problem. At times in the past, such as 55 million of years ago, there was a dramatic rise of about 5°C in the temperature of the surface layer of the oceans, and even more in the deep ocean. Sealevels rose by 5-6 metres as a result. The rate of addition of carbon into the atmosphere 55 million years ago was very high, but still less than the rates of today. However, geological research tells us how it all worked out, and the answer is not very reassuring. It took well over 100,000 years for Earth's climate system to return to the state that existed before the warming event. Faunas and floras on land and in the sea took a big hit, especially those organisms living on the bed of the ocean.

The cause of this 55 million year old event is the subject of scientific debate, but the majority of workers believe it was caused primarily by the release of methane from a frozen state in sediments beneath the ocean floor. Methane is a much more powerful agent in the greenhouse effect than carbon dioxide. Despite the fact that the trigger for the event was different to that implicated today, the 55 million year old event stands out as an example of the sort of pathway on which we may be embarked. Once it has been established that we have a problem, the best people to understand the climate system well enough to predict climate trends in the near future are physicists, chemists and computational scientists working on atmosphere, land and oceans.

Understanding the 55 million year ago event is made possible by an ability to construct a method of dating, or chronology, with a resolution of thousands of years. This is a stunning geological achievement, but doesn't quite compare with watching the mood swings of planet Earth on a daily, yearly or decadal time scale. Like anxious parents, we look at the latest graphs of carbon dioxide concentrations in the atmosphere, mean surface temperature, and extent of the world's ice caps; we are bombarded with news items on devastating floods and searing heat waves and wonder if we are witnessing the restlessness of Earth under anthropogenic climate change. Meanwhile, climate sceptics clutch at every apparent contra-indication. For example, global surface temperatures were rising rapidly during the final decades of the last century, but the temperature rise has stalled in the last 15 years[6]. The reason may be something to do with a reduction in the radiation received by the Earth from the Sun, which varies on a time scale of tens of years, but may also be due to a greater uptake of heat than expected by the oceans. Whatever the cause, in the perception of the sceptic, this is an example of the model being wrong, so why should anyone believe climate predictions at all?

Such is the challenge to communicate the essential (if not 'inconvenient') truths about the climate system and climate change to policy makers and publics that various position statements have been put out by learned societies. The most prominent (in the UK) are perhaps the Royal Society's (2007) *Climate Change Controversies: A Simple Guide*, and the

Geological Society's (2010) *Climate Change: Evidence from the Geological Record.*
Both documents seek to answer key questions that loom large in what is
often a fractious debate. The Royal Society's remit was the wider of the
two, structuring the information around 8 misleading arguments, but they
did not consider the value of information from the deep past. The
Geological Society's statement, instead, is focused on what geologists know,
which is based on rocks, sediments, and ice-cores rather than what we kid
ourselves is true from computer models. Together, these documents
provide a consensus view of a complicated field of science made relatively
simple for consumption by the literate but non-specialist reader. I think this
is unprecedented, and very valuable too.

What then, provides the oxygen for a continuing scepticism of
anthropogenic climate change? To answer this in part, we need to go back
to the oil industry that finds the carbon, gets it out of the ground, refines it
and sells it. Large oil companies are unlikely to wind up their businesses in
the face of a low carbon economy, so their attitudes are key to everyone's
future.

Bryan Lovell[7] writes on the changing outlooks of the oil industry
with the benefit of knowing the key players, and having worked with some
of them as colleagues. He highlights the divide of opinions either side of
the Atlantic, BP and Shell on the European side embracing climate change
as a new business opportunity at the same time as helping to solve an
acknowledged environmental problem, while in USA major oil companies
played the role of innocent sceptics. Critically, BP and Shell top
management was composed of, or heavily advised by geologists, and very
fine geologists at that. David Jenkins was Chief Geologist of BP, then
Technical Director, and advised Lord Browne, the CEO of BP, which led
to a change of direction by the company. Sir Mark Moody-Stuart performed
a similar role as chairman of Shell, acknowledging before the Kyoto
meeting of 1997 that there was now sufficiently credible scientific evidence
to support anthropogenic climate change.

Meanwhile, Stateside, the picture was very different. In an attempt
to coordinate opposition to the growing weight of opinion on
anthropogenic climate change, major oil companies worldwide joined the
USA-based Global Climate Coalition in 1989. It is now defunct, BP being

the first to withdraw in 1996, just prior to the public announcements of a change of attitude and eventual rebranding by BP. BP's conversion was more due to internal recognition than external pressure, but was made possible by the steadily accumulating weight of scientific evidence. Chief Executive Tony Hayward, also a geologist, was a firm believer in anthropogenic climate change and an optimist that BP could rise to the major challenge of contributing to bringing about a low carbon economy. BP, with its wealth of geologists in leadership positions, was therefore ahead of the game in its position on anthropogenic climate change.

In contrast, North American companies such as ExxonMobil were slow to respond. They continued to see no human influence on climate change, feeling by the end of the millennium that scientific understanding was still insufficient. Exxon and Mobil, then separate companies, opposed the Kyoto agreement of 1997 in anticipation of US president George Bush's formal opposition shortly later. Lee Raymond, Exxon Mobil's CEO then Chairman from 1993 to 2005, a chemical engineer, took a combative view on climate change, believing global warming to be a giant hoax. With the succession of Rex Tillerson as Chairman, an exploration and production engineer, ExxonMobil now (2013) take a softer line, acknowledging that climate change is real, and investing in research to allow it to develop low carbon technologies.

Much has therefore changed in the last decade. The American oil industry, and their mouthpiece, the American Association of Petroleum Geologists, is now positioning itself to play a major, constructive role in bringing about a low carbon economy. However, the American oil companies, together with other international giants, and the state-owned companies that control an increasing percentage of the world's so-far unexploited hydrocarbon reserves, need the help of national governments to make a success of it. The Carbon War? If there is a war, it is between a coalition of badly informed environmentalists dealing with propaganda and a cynical general public suspicious of anything delivered by politicians and business elites, arranged against a broad consensus within scientific and corporate organizations.

The Yale Climate Project (http://environment.yale.edu/climate-communication/) was formed in order to bridge the gap between science and society specifically in the area of climate change. This is a considerable undertaking. When surveyed in 2013, 70% of the inhabitants of Texas – perhaps the most pro-oil state in the USA – believed that global warming was taking place, but less than half (44%) believed that it was caused by human activities[8]. Contrast this with a declaration of the IPCC in September 2013 stating that the likelihood of the observed warming being driven by human activities was now 95%. In terms of public perception, therefore, many are in denial that man is changing global climate at all. To those who believe that climate change is taking place and is caused by man, the hydrocarbon industry is the villain. This presents a serious communication problem. Tempting as it is to regard these social, cultural and psychological aspects as 'barriers' to be swept aside in order to usher in a brave new world of public understanding, it is probably more the case that these social barriers are here to stay, and that policies for energy need to take account of them and use them. In other words, the problem isn't human ignorance, but the lack of understanding, and perhaps the lack of an interest in gaining an understanding, of how and why people think and act.

Research institutes are hard at work investigating how a reduction in carbon emissions might come about that would hold atmospheric concentrations at, say, 500 parts per million. Pacala & Socolow[9] wrote in 2004 that:

> 'Humanity already possesses the fundamental scientific, technical and industrial know-how to solve the carbon and climate problem for the next half-century'.

What is this striking assertion based on? It is based on 15 technologies and actions that together could hold atmospheric concentrations at just above present-day levels. Others believe this is wildly optimistic. One of the most important technologies is the capture and storage of carbon dioxide deep underground. Carbon dioxide can be taken from the flues of coal-fired power stations, or from that generated during the production of hydrocarbons from reservoirs, and be safely re-injected underground into porous rocks where it can be stored for very long time periods without danger of escape or of contaminating the groundwaters needed for agriculture. Carbon storage or sequestration is undergoing intensive investigation and makes use of the very same skills used by geoscientists and engineers in their conventional exploration and production activities. In my own institution, a consortium including Shell, Qatar Petroleum and Qatar Science and Technology Park, is funding a large research programme focused on the safe storage of carbon dioxide in carbonate rocks. Many other coalitions of research groups around the world are hard at work on this same topic.

As we have seen before in other contexts, environmentalists have voiced strong objections to carbon capture and storage. At first sight this is puzzling, since what could be negative about taking carbon dioxide from a coal-fired plant and storing it where it can do no harm? Carbon capture from coal-fired power stations in China and India, for example, could make a serious dent in those countries' carbon dioxide emissions. The environmentalist lobby takes the view that the safest strategy is to crack down on all activities that involve the burning of fossil fuels and to accelerate the development of renewable enrdgy sources. This is understandable as a policy position, but there are some awkward realities to be faced.

As an illustration, one of these realities is that the building of renewable infrastructures, such as wind farms, requires colossal amounts of raw materials, which must be dug from the ground, concentrated and manufactured, consuming large amounts of energy, which are provided by the burning of fossil fuels. Wind farms are therefore far from carbon neutral.

Whether it is carbon capture and storage, or the exploitation of shale gas by hydraulic fracturing, or anthropogenic climate change in general, the public have lost confidence in 'experts' and political leaders. The attempts by scientists to bridge this gap in trust have been clumsy, not giving due credence to uncertainties, making exaggerated claims, exacerbated by competition between research groups for funding and status. Having said this, there is remarkable consensus among a very wide range of climate scientists, as shown by the periodic statements of the Intergovernmental Panel on Climate Change. A slightly weaker consensus extends to those briefed by the scientific community, such as policy makers. A learning point from the climate change debate is that social 'barriers' and a general post-modernism that makes a virtue of diversity of opinion or outright denial are particularly strong factors at work. The lessons learnt from Earth history revealed by geologists have been successfully translated to the public and to policy makers in terms of the natural variability of change, but less so in terms of the well-documented analogues of a greenhouse world. If these analogues are allowed to set the intellectual context for anthropogenic global climate change, the field is clear for a wide range of climate scientists to make their quantitative predictions of a range of climatic variables for the next 50 years. Thus far, the ability of the geological community to act as an informed voice on climate change remains in doubt[10].

Shale Gas and Eco-warriors

Watching on the television the protests against shale gas exploration in the gloriously pastoral countryside of the South Downs of England left me puzzled. The extraction of gas from shale (a common sedimentary rock made of compressed very fine particles with tiny pores that may hold water or gas) by hydraulic fracturing has become the current focus of eco-warriors. Fracturing is required to provide pathways for the extraction of gas from the densely compacted shale rocks.

First, it needs to be said that there's a lot of it about. Shale is one of the most common sedimentary rocks, generally resulting from deposition of clay-sized particles (less than 4 microns in diameter) in the quiet conditions of the seabed. The process of extracting natural gas from shales has really taken off in USA, where in 2010, shale gas accounted for 20% of all natural gas production and that figure is likely to more than double by 2035. China probably has even larger reserves.

Shale gas has been extracted in USA since 1821, but the first enhancement of gas production by hydraulic fracturing took place in 1947. By 2013, thousands of wells have been drilled to hydraulically fracture shale gas reservoirs. This has secured a strategic supply of gas for the USA, helped its balance of payments, provided cheap gas for US consumers, and, importantly in this context, brought CO_2 emissions in 2012 down to a 20 year low, helped by the recession, which ramped down industrial production. Use of shale gas lowers CO_2 emissions as long as it replaces the burning of coal. However, it emits more methane than conventional natural gas, though less than coal.

Why were eco-warriors out in such force in 2013 at Balcombe in the South Downs? Their concerns appear to be two-fold. The first set of concerns includes the leakage of gases, possible contamination of groundwater supplies, above-ground problems of waste water containing unwanted chemicals, and earthquakes caused by hydraulic fracturing. Organizations such as Greenpeace cite the risk of polluting water supplies, and have encouraged farmers to deny access rights to drilling companies. Fears have been swelled by the 2010 documentary *Gasland*, which focused on negative impacts. Yet a study at Massachusetts Institute of Technology in 2011 stated:

'The environmental impacts of shale development are challenging but manageable'.

Regarding earthquakes, the waves produced during fracturing are microseismic events below the detection levels of humans. Rare occurrences that were felt by humans were far too small to cause any damage. A US government report in 2012 stated[11]:

'About 35,000 hydraulically fractured shale gas wells exist in the United States. Only one case of felt seismicity in the United States has been described in which hydraulic fracturing for shale gas development is suspected, but not confirmed. Globally only one case of felt induced seismicity at Blackpool, England has been confirmed as being caused by hydraulic fracturing for shale gas development'.

The second set of concerns of the environmentalist lobby is about the contribution that the burning of shale gas will make to greenhouse gas emissions. If shale gas replaces coal in the energy mix of an economy, there is a reduction in the CO_2 emissions for a unit of energy produced. If the energy mix is dominated by nuclear, as in France, the adding of shale gas to the energy mix will *increase* that country's greenhouse gas emissions. Shale gas is therefore good or bad news depending on what it replaces in the global energy mix. From an absolutist point of view, few would argue with the aspiration to cut greenhouse gas emissions, but can this be done in a managed way?

Environmental organizations such as Greenpeace question the benefits of reduced CO_2 emissions by the exploitation of shale gas, and selectively cite reports or paragraphs and sentences of reports that support their sceptical stance. They do not regard shale gas as a means of keeping the lights on while a renewables-based low carbon economy is progressively developed.

Carbon Capture and Storage

As introduced above, carbon capture and storage (CCS) is the putting back of carbon deep underground where it came from[12]. Carbon dioxide is produced as a waste product during the separation of hydrogen from natural gas and from the burning of fossil fuels in power plants. Separating a compound from a mixture gets increasingly difficult as the concentration of the compound decreases, which is why it would be fantastically expensive to extract CO_2 from the air of the atmosphere. Separating it by the various industrial processes at a power station reduces the overall efficiency of the plant, but yields low-carbon electricity from coal, gas and biomass starting materials. The compressed and dehydrated CO_2 would then be transported by pipeline for underground storage. Consequently, a key factor in the economics of CCS is the proximity of a power plant to the oil field from which the oil and gas are extracted and in which the CO_2 will be stored safely. However, injection points and storage sites of CO_2 must be chosen carefully to avoid interference with existing oil and gas operations.

At the rapid rate at which research on CCS is progressing, the technical challenges will probably be solved in the near future. However, whether this will be achievable economically, and at such a scale as to make a dent in global greenhouse gas emissions is less certain. The high cost end of the process is the capture of carbon dioxide, involving retro-fitting of existing power stations for capture after combustion, or the building of new power stations to capture before combustion.

The volumes of CO_2 required to be captured and stored in order to make a real impact on greenhouse gas emissions is daunting. The capture and storage of just 4% of global emissions of carbon is equivalent to about a third of the current flow rate of oil from underground reservoirs[13]. It is unlikely that there is sufficient underground storage potential in existing oil and gas fields to keep carbon emissions from China's and India's coal based power stations out of harm's way.

The problem is not strictly the availability of pore space in underground reservoirs, since carbon dioxide could be stored in water-filled aquifers as well as in oil and gas fields. However, the advantage of using oil and gas fields is that we know that hydrocarbon fluids have been stored there for millions of years, because we have extracted some of them, whereas aquifers may be leaky and unable to contain carbon dioxide at high pressure. Consequently, there is much research into the use of saline aquifers[14], since these provide the very large volumes of underground storage required.

Additional possibilities exist for trapping CO_2 by its reaction with the host rock, which therefore removes the need for a sealed reservoir. For example, it is estimated that a gigatonne of CO_2 per year might be sequestered by reaction with the rock peridotite to produce a carbonate[15]. Peridotite is rare at the Earth's surface, but is by far the most common rock in the interior of the Earth. We only see peridotite at the surface where tectonic movements have pushed slices of Earth's mantle tens of kilometres upwards. A further bonus is achieved if the injected CO_2 forces out increased volumes of hydrocarbons in what is called enhanced oil recovery.

The opposition of environmentalist groups to carbon capture and storage is in some respects understandable, since the capture process of CO_2 from power stations reduces carbon emissions drastically, but leads to an overall reduction of air quality. However, this is not what environmental lobbyists object to. They believe that CCS is a means by which the continued burning of additional fossil fuels will be justified. They state that the sheer volumes of liquefied carbon dioxide that would need to be stored underground are enormous. They also point to the reduction in the efficiency of power stations fitted to capture carbon, which would require more fuel for the same amount of electricity. This is also a genuine challenge. A concern is that money that might be invested in renewable energy sources will be diverted to CCS.

It was something of a surprise to those working on CCS projects when the European Commissioner for Climate Action, a new post in the

European Commission created in 2010, awarded money to renewables projects across Europe, but froze out CCS projects. This doesn't mean that CCS in Europe is dead, but much needs to be done to make it a viable proposition, especially in terms of who is going to pay for it. Meanwhile, sponsors of CCS projects are being forced into extreme risk aversion because of public attitudes. Together with other initiatives, including renewable energy, CCS has a potentially valuable role to play in satisfying energy needs so that the lights don't go out when changing to a low carbon economy.

The Carbon Crisis is public enemy number 1, and the industries providing the raw materials to generate energy are seen by many to be the great sinners. It is a peculiar state of affairs when the developed world gorges itself with overconsumption of raw materials, travel for business and leisure is booming and we are overwhelmed by electronic gadgetry of every kind, yet the sinners are apparently the energy and mining companies that satisfy these demands. The elephant in the room is that the simple answer to anthropogenic climate change is to reduce consumption.

Bearing in mind the seriousness of greenhouse gas emissions for future climate change, it is a sobering thought that the many pronouncements of impending doom are not matched with much government action or public alarm. There are several fault lines, between policy makers, scientists, environmentalists and scientists. The great challenge is of course obvious – to reduce the impact of future climate change while avoiding the lights going out. To achieve that challenge will require more than shrill propaganda. It will certainly require the participation of technologically advanced partners. Sinners will by degree need to become saints.

13 EPILOGUE: DARK SARCASM IN THE CLASSROOM

In which truth is in the witness box, and a plea is made for science to be humanized . . . .

Is truth 'what your contemporaries allow you to get away with'?[1] Practicing science as a career and as a passion, most scientists are inclined to think that apart from being a lot of fun, it is worthwhile. If truth is not in some small way capable of being discovered, then it becomes solely a construction of the human mind. In other words, you do not 'discover' truths but you 'construct' them.

Do things exist in the natural world if they are not being seen, or perceived? Take colour. The human eye causes colour to be perceived by the brain by breaking down the different wavelengths of light that is received by the retina. But if you stop looking, then colour doesn't exist. Objects do not possess colour independently of the observer; it is an image, not a copy of the external world. So it is with science. We *create* a view of the world in the terms of science and mathematics (or alternatively in the terms of art, sculpture, poetry, literature) but it is through a creative act of the human mind, and is not a photographic copy.

Despite this, and jumping acrobatically over vast mine-fields of philosophical thought, one of the central tenets of science is that there is an objective truth to be discovered, and that the truth is not changeable or an illusion, no matter how many postmodern philosophers are lined up to say differently. This objective truth includes that light consists of photons with wave-particle duality, even if colour is another thing. Science therefore confidently makes a claim for itself that it is following a path to objective truth. This path, however, is tortuous. Just as science, perhaps rightly, takes up an elitist position in society on the pedestal of knowledge, it suffers the fate of all elites, a dark sarcasm in the classroom[2].

Geology deals with how the Earth and planets behave today, but has the added ingredient of time, and lots of it. This paints the complexion of geology as seen by the outside world, but also runs through the arteries of the discipline. It is manifested in the attitudes that geologists take to the many problems that occupy them. As an insight, Rachel Carson (of *Silent Spring* fame) wrote that sediments are the epic poem of Earth history. The long-term resting place for sediments is what we call stratigraphy, since 'strata' are layers, and sedimentary rocks appear to be layered at various scales. Stratigraphy and the long history of this planet that its study reveals, is, however, a little like two upward tapering lines in a picture – indecipherable as the apex region of a triangle, or as two equally spaced lines disappearing into the distance like rail tracks. It is a question of perception. There are reasons for this.

First, stratigraphical records, like historical records, are a partial, filtered record of all that happened. Second, the way the world functions today is no more an indication of how it functioned in the past, than it is to say that the 21st century is historically a good indicator of the 1st. The present may well be the key to the past, as the Lyellian mantra goes, in the sense that universal laws have operated since the beginning to the present day. There was no new physics or new chemistry that took over, no new biology that governed Earth's intricate co-evolutionary pathway. New organisms were spawned and extinguished, new minerals were formed in a chemical evolution, continents aggregated, dispersed and rejoined, but Earth was lapt in a universal law. If this is what 'the present is the key to the past' means, then we have no problem with it. But we struggle to use the present to understand the geological record of the past because there are things that have happened over the vast expanse of geological time that have never been recorded, let alone experienced, by Man. We are restricted by the short time-scale of direct observation. 'Seeing' the present day leaves us a long way short of being equipped to read the epic poem. 'Seeing' the present-day is essential, but is entirely inadequate to read Earth's great narrative of the past.

The intricasies of the trajectory of science and the inherent flaws in this very human system are a product of the world-pictures of those who practice it. Having these world-pictures outside in the light of day for all to see is admittedly rather like hanging our one's dirty linen. It is a purging though chilling process that one day might result in less dark sarcasm in the classroom. Indeed, one can be heartened by the 'humanization' of science, since scientific enquiry is not only about distant galaxies. It is about the dynamic and dangerous world that we humans inhabit, and is carried out by humans with joys and sadnesses, successes and failures, hopes and fears.

The medieval scholar Nahmanides stated eight hundred years ago that:

'the world [channelled by the laws of Nature] functions according to its natural pattern'[3]

and it is this natural pattern, of which humans are a part, that scientists seek to understand. The physicist John Polkinghorne puts it poetically[4]:

'It is in the nature of dense snowfields that they will sometimes slip with the destructive force of an avalanche. It is the nature of lions that they will seek their prey. It is the nature of cells that they will mutate, sometimes producing new forms of life, sometimes grievous disabilities, sometimes cancers. It is the nature of humankind that sometimes people will act with selfless generosity but sometimes with murderous selfishness.'

The question of the role of Nature in the affairs of Man is not a new one. It begins in an era of widespread religious belief in a God that was neither distantly indifferent nor acted like a cosmic tyrant in a puppet theatre[5]. The idea of historical events, some natural, as a form of divine revelation was part of common thinking in the ancient Near East and Hellenistic world. What has to be asked is whether an event was revelatory in itself, or became revelatory as a result of its interpretation. One such event took place in Lisbon, Portugal in the year 1755. The event was one of the strongest earthquakes in recorded history, reaching 9 on the Richter scale. This means that the earthquake released over a hundred times more energy than the 1999 earthquake in northern Turkey and one thousand times more energy than the 2009 earthquake of L'Aquila, central Italy. Not surprisingly, the earthquake shocks destroyed this wealthy city, with 50,000 to 60,000 fatalities. The earthquake brought down buildings, caused extensive fires and to add insult to injury, generated a tsunami that struck the city as a wave over 15 m high. The timing of the earthquake was seen by many to be extremely significant, on All Saints Day. It was therefore seen as an example of divine retribution, of the result of God's anger.

The Lisbon earthquake blew apart the idea that man and Nature were at peace, a view particularly associated with Jean-Jacques Rousseau (1712-1778) but echoed by romantic poets such as William Wordsworth. Rousseau believed in the original goodness of human nature in the 'noble savage'. He believed that the closer a man was to Nature the more virtuous he became. Voltaire (1694-1778) strove to counter what he saw as false optimism. Although he believed in God, he was violently opposed to the Roman Catholic Church. He advocated a harsh sense of realism and championed the idea of man tormented by a Nature out of control. He argued that an all-powerful God could have prevented the Lisbon earthquake, and the fact that he did not, demonstrated that there was no benevolent protection for the righteous. Voltaire's 'Le mal est sur la terre' replaced Rousseau's idea that 'tout est bien'.

The reactions to the Lisbon earthquake and the questions surrounding the role of God in Nature have a modern ring to them. They are rooted in the Deep Ecology movement, though God has been replaced by a green version of mysticism. Doing geological fieldwork in stupendously beautiful places like the national parks of Utah, the beauty of Nature is stunning, awesome, and somehow spiritual. I picked up a copy of Bill McKibben's book *The End of Nature* [6] from a bookstall on the edge of Zion National Park in southwestern Utah. Many people would echo McKibben when he admits that he finds a certain spirituality located in Nature. Yet what we sense when we see the beauty of Nature is perhaps not contained in Nature itself, but instead reflects our yearning for a beauty that we see through or beyond Nature [7]. In the same way, we are moved by music, poetry, art or books, not by the beauty contained in them, but by a longing for beauty that rises up in us when we listen to or read them. What we are longing for is not the thing itself, whether unspoilt Nature, music, poetry or books. Quoting C.S. Lewis (1941), these things are only

'the scent of a flower we have not found
the echo of a tune we have not heard,
news from a country we have not visited'.

We may see beauty through Nature, but this same Nature convulses and takes lives by the thousand. We cannot have one without the other, in fifty shades of Rousseau and Voltaire. It is this world that geological scientists are interested in and have argued about for centuries.

In a world in which science has achieved a transcendent status, there is an understandable triumphalism that veers into arrogance. Communication of scientific discoveries to the various publics is frequently bland and one-dimensional. It is too often forgotten by scientists that the publics whose approval they seek are human, and that the scientific world in which they operate is also intensely human. The world is comprehensible to science, which is in itself a remarkable thing, but that does not mean that science is the sole language for experience; it merely applies brush strokes from human hands using the palette of colours available.

'Science, like art, is not a copy of nature but a re-creation of her'[8].

Our overview of a number of ancient mysteries and modern controversies from the field of geology and environment merely scrapes the surface in revealing the human factors at work in the progression of the scientific enterprise and demonstrates the over-riding importance of world-pictures. If scientists and publics alike could realize that the scientist merely paints his or her picture of reality, it would be a lot easier to translate exciting scientific discoveries effectively.

Anyone who provides answers is not properly a scientist. A scientist only generates questions.

ENDING

The light was fading when I ventured out to collect dry birch logs for the
fire.
I noticed that the clouds of Woodstock were no longer weeping.
Smoke from the chimney struggled in its upward ascent into a clear sky.
The land seemed to fall silent as the sky grew imperceptibly darker,
and the evening air took on a delicate fragrance.
As in a tired child, rubbing its eyes, thoughts subsided.
Deeper senses were aroused,
senses of regret and grieving and thankfulness in equal measure.
Two silhouetted shapes fluttered across the rooftop,
then rose in unison, became smaller, and soon were gone.

NOTES

Preface
[1] The subtitle of Robert Park's book *Voodoo Science*, Oxford University Press, 2000, 230 pages.
[2] A question posed by Joan Fujimura, 1991, On methods, ontologies, and representation in the sociology of science: where do we stand? p.207-248 in *Social Organization and Social Process: Essays in Honor of Anselm Strauss*, ed. by D. Maines, New York, Aldine de Gruyter, page 222.
[3] Peter Medawar, *The Limits of Science*, 1984, Oxford University Press, 108 pages.

Chapter 1 Making Dramas: the Secret Workings of Science
[1]*Genesis*, ch 15, verse 5: He took him outside and said 'Look up at the heavens and count the stars – if indeed you can count them'. Then he said to him, 'So shall your offspring be'. *Holy Bible, New International Version*. Actually, the unaided human eye can see about 2500 stars in the night sky, but in our galaxy alone, the Milky Way, there are thought to be 100 billion stars.
[2] Barnes, B., Bloor, D. & Henry, J., 1996, *Scientific Knowledge: A Sociological Analysis*. University of Chicago Press, Chicago, Chapter 2, 230 pages.
[3] Mary Midgeley, 1985, *Evolution as a Religion: Strange Hopes and Stranger Fears*, Methuen & Co. Ltd., London, 180 pages.
[4] Based on M. Heidegger's *Being and Time*, 1962, In: Macquarrie, J., Robinson, eds., Harper and Row, New York, 589 pages. Translation of the German original of 1927.
[5] This idea of a social contract comes from Jean-Jacques Rousseau, 1762, *Du Contrat Social*, though antecedents are found in Greek philosophy.
[6] Jacob Bronowski, *Science and Human Values*, p.10, Harper & Row, New York, 1972, 119 pages.
[7] Francis Galton (1822-1911) was a Victorian polymath, who introduced the phrase 'nature versus nurture'. His ideas on eugenics are set out in *Hereditary Genius: An Inquiry into its its Laws and Consequences*, first published 1869 by MacMillan, London, 390 pages.
[8] Walter Gratzer, *The Undergrowth of Science*, p.299, Oxford University Press, 328 pages.
[9] An unusual form of water was 'discovered' by the Soviet chemist Boris Derjaguin, and reported in an epidemic of papers beginning in 1962, but its anomalous properties were due to impurities of silica from laboratory glass. Chapter 4, Gratzer *op cit*.
[10] Irving Langmuir (1881-1957) coined the term 'pathological science' in 1953, and it was adopted by Walter Gratzer in *The Undergrowth of Science, op cit*. Robert Park included pathological science, junk science, pseudoscience and fraudulent science under the term 'Voodoo Science' (Robert L. Park, *Voodoo Science: the Road from Foolishness to Fraud*, Oxford University Press, 2000, 230 pages).
[11] Underdeterminism is a concept in the philosophy of science that refers to situations where the available evidence can be interpreted in different ways and does not point to a single conclusion. See, for example, Stanford, Kyle, 2000,

Underdetermination of Scientific Theory, Edward N. Zalta, ed., The Stanford Encyclopedia of Philosophy.

[12] From A.D. Miall & C.E. Miall, 2001, p.323, citing Frodeman, R., 1995, Geological reasoning: Geology as an interpretive and historical science. Geological Society of America, Bulletin vol.107, p.960-968, and Dott, R.H. Jr., 1998, What is unique about geological reasoning? GSA Today vol.10, p.15-18.

[13] Robert L. Park, *Voodoo Science: the Road from Foolishness to Fraud*, Oxford University Press, 230 pages.

[14] Charles Babbage (1791-1871) was an English polymath who, amongst other things, invented the idea of a programmable computer. He sharply criticizes science in *Reflections on the Decline of Science in England and some of its Causes* (published by B. Fellowes, London) in 1830, whose publication led to the formation of the British Association for the Advancement of Science in 1831.

[15] The term 'sceptered isle' comes from William Shakespeare's Richard III, "This royal throne of kings, this scepter'd isle, This earth of majesty, this seat of Mars ... This blessed plot, this earth, this realm, this England", since the finding of early human remains at Piltdown gave England its own artifacts to rival those discovered in France and Germany.

[16] Archaeologist Miles Russell of Bournemouth University, England, *Piltdown Man: The Secret Life of Charles Dawson*, Tempus, 2003, 288 pages.

[17] Stephen Jay Gould (1980) thought Teilhard de Chardin was the culprit. We meet him again in Chapter 5.

[18] Langmuir's stigmata are taken from Walter Gratzer's *Undergrowth of Science*, op cit., p.79-80.

[19] Seven Warning Signs of Bogus Science, Robert L. Park, The Chronicle of Higher Education, Jan 31, 2003.

[20] www.mantleplumes.org, Zombie Science and Geoscience, D.L. Anderson and Warren B. Hamilton. Last accessed 10 December 2013.

[21] Armstrong, R.L., The persistent myth of crustal growth, Australian Journal of Earth Sciences, vol.38, p.613–630, 1991.

[22] Walter Gratzer, *The Undergrowth of Science*, p.299, *op cit.*

[23] Edward Osborne Wilson, *Consilience: The Unity of Knowledge*, 1998. Consilience is here used to mean the synthesis of knowledge from different fields of specialization. E.O. Wilson was particularly thinking of the linking together of physical and life sciences with social sciences and humanities.

[24] Bryan Lovell states in *Challenged by Carbon* (2010, Cambridge University Press, 212 pages) that 'you can't argue with a rock'. A similar sentiment is found in the title of David Montgomey's book *The Rocks Don't Lie: A Geologist Investigates Noah's Flood*, W.W. Norton, New York, 2013, 302 pages.

[25] G. Bowker 1994, p.157, based on a study of Schlumberger, *Science on the Run: Information management and industrial geophysics at Schlumberger 1920-1940*, Cambridge, MIT Press.

[26] The idea of social worlds is elaborated by Adele Clark, 1991, *op cit.* p.131.

[27] Induction is the derivation of beliefs from the assembly of many disparate observations or experiences. Francis Bacon (1561-1626) was a strong advocate of this approach.

[28] Robert L. Park, 2000, *Voodoo Science: The Road from Foolishness to Fraud* p.40, Oxford University Press, 230 pages.

[29] *The Limits to Science*, Peter Medawar, 1984, Harper Collins, 108 pages.

[30] From Alister E. McGrath, 2011, *Surprised by Meaning: Science, faith and how we make sense of things*, p.40, Westminster John Knox Press, Louisville, Kentucky, 136 pages.

[31] Steve Jones, *The Serpent's Promise*, p.15, Little, Brown Publishers, London, 2013, 437 pages.

[32] T.H. Huxley, *Collected Essays*, vol. 4, p.139-163, 1895, London, Macmillan.

[33] Webster, R. G. and Walker, E. J., 2003, American Scientist vol.91(2), p.122. doi:10.1511/2003.2.122.

[34] http://izquotes.com/author/ernest-rutherford, last accessed 10 December 2013. Quoted in *Einstein: The Man and His Achievement*, 1973, by G. J. Whitrow, p.42.

Chapter 2 Shifting Foundations: Continents Adrift

[1] Wegener in fact matched the 200 m depth contour rather than the coastline, since he was aware that the coastlines may have changed by the processes of erosion and the deposition of sediment by large rivers.

[2] Antonio Snider-Pellegrini, *La Création et ses Mystères Dévoilés* – 'The Creation and its Mysteries Unveiled', 1858.

[3] William Whewell, *The Philosophy of the Inductive Sciences* (2 volumes, John W. Parker, London, 1847), vol.2, p.36.

[4] Jacob Bronowski, p.12 and p.14, *Science and Human Values*, 1972, Harper & Row, New York, 119 pages.

[5] From Edna St. Vincent Millay's *Collected Sonnets*, rev., exp. ed., Harper Perennial, New York, 1988, p.140. Quoted in Alister E. McGrath, *Surprised by Meaning*, p.103, Westminster John Knox Press, Louisville, Kentucky, 2011, 136 pages.

[6] Ted Nield, *Supercontinent*, Granta Press, especially chapter 7, World Wars, p.125-156, 288 pages.

[7] Mott Greene (Professor of History of Science at University of Puget Sound) believes that much has been made of the fact that Wegener was not a proper geophysicist but an 'interloper'. Quoted in T. Nield, *Supercontinent*, p.135.

[8] Ted Nield, p.135, *Supercontinent*, *op cit*.

[9] Curie, P. and Laborde, A., 1903, Sur la chaleur dégagée spontanément par les sels de radium, Comptes Rendus de l'Académie des Sciences, vol.136, p.673–675.

[10] E. Rutherford and F. Soddy, 1902, Philosophical Magazine 4, p.370-396.

[11] Vladimir Beloussov, *Basic Problems of Geotectonics* (translation), 1962.

[12] As Naomi Oreskes (Professor of History and Science Studies at San Diego) suggests, *The Rejection of Continental Drift: Theory and Method in American Earth Science*, Oxford University Press USA, 1999, 432 pages.

[13] Quoted in Rutherford at Manchester, 1962, by J. B. Birks, http://izquotes.com/author/ernest-rutherford.

Chapter 3 Victorian Myths and Dirges about Nature

[1] For example, the influential work by I. G. Barbour, 1966, *Issues in Science and Religion*, SCM Press, and J. Polkinghorne, 1994, *Science and Christian Belief: Theological Reflections of a Bottom-Up Thinker*, SPCK.

[2] As reported in *The Guardian* newspaper on the unveiling of a plinth in Oxford to celebrate 150 years since the Wilberforce-Huxley debate, Alison Flood (10 September 2010), "Plinth commemorates Huxley-Wilberforce evolution debate".

[3] The review can be found at 'The Victorian Web', www.victorianweb.org/science/science_texts/wilberforce.htm. The sheer length and detail of this review is noteworthy. It appeared in Quarterly Review, 1860, p.225-264.

[4] J.R. Lucas, 1979, Wilberforce and Huxley: A legendary encounter. *Historical Journal* 22, p.313-330.

[5] Suggested in Ruse, Michael, 2001, *Can a Darwinian be a Christian? The Relationship between Science and Religion*, Cambridge University Press. p.5. ISBN 0-521-63716-3.

[6] Quoted in Mary Midgley's *Evolution as a Religion* (Methuen, London), 1985, p.11, but derived from J.R. Lucas, Wilberforce and Huxley: a legendary encounter, published in The Historical Journal vol.22, 2 (1979). More background to this fascinating dispute can be found in Gilley & Loades, 1981, Thomas Henry Huxley; the war between science and religion, published in The Journal of Religion vol.61, 3.

[7] Huxley coined the term 'agnostic' and allegedly professed to be one, and his reputation emphasizes his opposition to the intrusion of religion into explanations of the natural world. Nevertheless, he never rejected the argument that evolution operated in a world created by God, and was therefore a deist-evolutionist. Huxley did not disagree with Wilberforce on the basis of the existence of God. Instead, he was stoutly defending the idea of evolution from religious interventions, which is quite different.

[8] *Vanity Fair*, July 24, 1869, the year of publication of Charles Darwin's 5th edition of *Origin of Species*.

[9] In one phrase I have glossed over an immense field of debate as to whether the story of the evolution of living things on Earth has been by gradualistic change or by punctuated steps. The differences between gradualistic and punctuated models is discussed by Stephen Jay Gould and Niles Eldredge in 1972 in Eldredge, N. and Gould, S.J., 1972, in T.J.M. Schopf, editors, *Models in Paleobiology*, San Francisco: Freeman Cooper, p.82-115. Reprinted in N. Eldredge, *Time Frames*, Princeton University Press, 1985, p.193-223. The idea of punctuated equilibrium was put forward by Stephen Jay Gould & Niles Eldredge, 1977, Punctuated equilibria: the tempo and mode of evolution reconsidered, Paleobiology vol.3 (2), p.115-151.

[10] Most writers would use the phrase 'struggle for survival' rather than 'game of life', but the former perhaps gives the wrong impression. Individual organisms may live to maturity and produce healthy offspring that themselves reproduce. Other individuals in a population may die before they are able to reproduce, so do not pass on their inherited genetic chemistry to offspring. Acting in a population over time, the genetic characteristics of the fecund individuals become dominant. So who is struggling to survive? Yes, there is life, reproduction and death, but there is no particular struggle for survival for individuals. 'Game of Life' is a phrase used by Per Bak in *How Nature Works*, 1996, Springer, New York.

[11] J.C. Greene, *Debating Darwin: Adventures of a Scholar*, Regina Books, 1999, 288 pages.

[12] The Chernobyl nuclear plant accident in the Ukraine in 1986 released large quantities of radioactive particles into the atmosphere. The main health impact was thyroid cancer in children. Long-term genetic defects may result from the accident, as shown by a doubling of the rate of certain damaging mutations among children born after the event.

[13] Suggested by geneticist Steve Jones in 2008-2009, on the basis that firstly, in today's comfortable world, genetic mutations fail to confer any significant advantage on the individual, secondly the rate of production of offspring has reduced, and thirdly because of globalization of human populations. William Stephens (*New York Times*, March 14, 1995) stated 'Natural selection has to some extent been repealed'. Steve Jones gave public lectures entitled 'Is human evolution over?' and radio interviews on the same subject but there appears to be little published in conventional scientific journals. The view has been strongly contested, but again in the media rather than in conventional scientific outlets. The statement is therefore controversial and unproven.

[14] L. Van Valen, 1973, A new evolutionary law. Evolutionary Theory, vol.1, p.1-30.

[15] From F. Heylighen, Principia Cybernetica Web, http://pespmc1.vub.ac.be/, created August 1993, 'The Red Queen Principle', last accessed 19 December 2013.

[16] A biological or evolutionary arms race was popularized by Richard Dawkins in *The Blind Watchmaker*, 1991, Penguin, 340 pages, soon after the publication of Dawkins, R. & Krebs, J.R. in 1979 of 'Arms races between and within species', in Proceedings of the Royal Society of London, vol.B 205, p.489-511.

[17] Ken Weiss, "Nature, Red in Tooth and Claw", So What? Crotchets & Quiddities, *Evolutionary Anthropology*, vol.19, p.41-45 (2010).

[18] A.C. Bradley's *A Commentary on Tennyson's 'In Memoriam'* (Second edition, revised 1907, The Macmillan Company, New York) is valuable, but I take full responsibility for the modern lingo.

[19] Michael T. Ghiselin, 1974, *The Economy of Nature and the Evolution of Sex*, University of California Press, 358 pages.

[20] A prospect discussed by the biologist John Maynard Smith in the New York Times Review of Books, 14 May 1992, p.34-6, 'Taking a Chance on Evolution'. The idea of repeatedly rerunning the tape of life comes from Stephen Jay Gould in *Wonderful Life: The Burgess Shale and the Nature of History*, 1989, W.W. Norton, New York, 347 pages.

[21] These are the contrasting views of Stephen Jay Gould (*Wonderful Life: The Burgess Shale and the Nature of History, op cit.*) and Simon Conway Morris (*Life's Solution: Inevitable Humans in a Lonely Universe*, Cambridge University Press, 2003, 464 pages).

[22] This view is held uncompromisingly by Richard Dawkins in *The God Delusion*, 2006, Bantam Books, 464 pages, and by Daniel Dennett in *Darwin's Dangerous Idea*, 1996, Simon & Schuster, 586 pages.

[23] Stephen Jay Gould, *Wonderful Life: The Burgess Shale and the Nature of History*, 1989, *op cit.*

[24] John C. Greene, *Debating Darwin: Adventures of a Scholar*, 1999, Regina Books, Claremont, California, 288 pages.

[25] *Life of Sir Isaac Newton*, David Brewster, 1875, revised and edited W.T. Lynn, new edition (Wlliam Tegg, London), p.303.

[26] Alister E. McGrath, *Surprised by Meaning*, p.74, Westminster John Knox Press, Louisville Kentucky, 2011, 136 pages.

[27] This view is stridently held by Richard Dawkins, *Darwin Triumphant: Darwinism as Universal Truth*, in *A Devil's Chaplain: Selected Writings*, 2003, p.78-90, Weidenfield & Nicolson, London, 264 pages.

[28] Page 325, Simon Conway Morris, *Life's Solution: Inevitable Humans in a Lonely Universe*, 2003, Cambridge University Press, 464 pages.

Chapter 4 Gaia: the Green Goddess

[1] C.E. Rosenberg, 1979, p.144, *No Other Gods*. Baltimore, MD, Johns Hopkins University Press, Baltimore, 366 pages.

[2] See Mary Midgley, *Evolution as a Religion*, especially Chapter 5, p.36-47, Methuen & Co. Ltd., 180 pages.

[3] Quote from Theodosius Dobzhansky, *The Biology of Ultimate Concern*, 1971, Fontana, London, 152 pages.

[4] Lovelock, James, 2001, *Homage to Gaia: The Life of an Independent Scientist* (Oxford University Press), 396 pages.

[5] Omega Point, the apex of human consciousness, was coined by the French geologist turned Jesuit priest Teilhard de Chardin in *The Future of Man*, 1950, reprinted 2004 by Image Publishers, 336 pages.

[6] Terry Eagleton, *Reason, Faith and Revolution: Reflections on the God Debate*, New Haven, Connecticut, Yale University Press, p.28, 2009, 200 pages.

[7] James E. Lovelock, *Gaia; A New Look at the Earth*, Oxford University Press, London, 1974, 176 pages.

[8] Harding, Stephan. *Animate Earth: Science, Intuition and Gaia*. Green Books, 2006, p. 44, 288 pages.

[9] Earth System Science is the investigation of how integrated physical, chemical and biological systems on the Earth work, both as a response to external forcing and through internal feedbacks.

[10] Watson, A.J. & Lovelock, J.E., 1983, "Biological homeostasis of the global environment: the parable of Daisyworld". Tellus vol.35B (4), p.286–9. Note the choice of language in 'parable', which echoes its meaning of simplified stories used to build understanding in faith-based communities such as Christianity.

[11] See for instance Kirchner, James W., 2003, "The Gaia Hypothesis: Conjectures and Refutations". Climatic Change vol.58 (1–2), p.21–45.

[12] M.E. Raymo and W.F. Ruddiman, 1992, Tectonic forcing of late Cenozoic climate. Nature, vol.359, p.117-122.

[13] Alexander von Humboldt (1769-1859) was a naturalist, geographer and explorer, who made early discoveries of the co-evolution of life.

[14] The Anthopocene, or Age of Man, has been suggested as a new geological epoch. The idea was nurtured by Paul Crutzen, the discoverer of the ozone hole. It is related, but in a complex way, to the idea of the Noösphere, or sphere of human cognition advanced by Vernadsky.

[15] James Lovelock published several papers on Gaia from 1975 onwards, but the most popular are *Gaia: A New Look at Life on Earth* (1979, Oxford University

Press), *The Ages of Gaia* (1995, Oxford University Press,) and *Homage to Gaia: The Life of an Independent Scientist* (2000, Oxford University Press). His more recent books have taken on a shrill tone of impending disaster, such as *The Vanishing face of Gaia: A Final Warning: Enjoy it while you can* (2009, Allen Lane Publishers).

[16] Margulis, Lynn. *Symbiotic Planet: A New Look At Evolution.* Houston, Basic Book, 1999, 154 pages.

[17] James W Kirchner, The Gaia Hypothesis: Conjectures and Refutations. Climate Change, vol.58, p.21-45, 2003.

[18] The term 'deep ecology' was coined by the Norwegian philosopher Arne Naess in 1973. Næss, Arne, 1973, 'The Shallow and the Deep, Long-Range Ecology Movement.' Inquiry, v.16, p.95-100.

[19] *Leaving Eden: To Protect and Manage the Earth* is the title of the book by Euan G. Nisbet, Cambridge University Press, 1991, 358 pages.

[20] From the political-social allegorical novel *Animal Farm* by George Orwell, published 1945, in which the animals chant 'Four legs good, two legs bad'.

[21] http://www.schumachercollege.org.uk/learning-resources/from-gaia-theory-to-deep-ecology. Schumacher College is named after the British economist E.F. Schumacher, the author of *Small is Beautiful: A Study of Economics as if People Mattered* (1973, Vintage Publishing, 288 pages).

[22] William Blake wrote these words in his preface to 'Milton a Poem', 1804, commonly interpreted to mean a breakdown of Nature and human relationships.

[23] This view is taken by the American historian Lynn White who delivered a paper in 1966 entitled 'The Historical Roots of the Modern Environmental Crisis'.

[24] The Select Committee sat in 2000, and the report was published by Demos in 2004.

Chapter 5 Paradise Lost? The Age of Man

[1] See http://ngm.nationalgeographic.com/2011/03/age-of-man/kolbert-text.

[2] Introduced by Teilhard de Chardin in 1922 and later developed by Vladimir Vernadsky, Noösphere means the sphere of human thought, from the Greek 'nous' meaning 'mind'.

[3] Derives from the concept of the Omega Point, the apex of human consciousness, which was coined by the French geologist turned Jesuit priest Teilhard de Chardin in *The Future of Man* (1950). Omega Man is a persistent theme in Mary Midgley's *Evolution as a Religion* (1985).

[4] A state of Omega Man, in benevolent control of a world in the Anthropocene, is echoed in the strident statement of Robert Park in *Voodoo Science* (p43) 'During the three and a half billion years of evolution, the environment shaped our genes. Our genes are now shaping the environment.'

[5] Reported by Crutzen, P. J., 2002, Geology of mankind. Nature, vol.415 (6867), p.23.

[6] Antonio Stoppani, *Corso di Geologia,* translated by Valeria Federighi, edited by Valeria Federighi and Etienne Turpin (Miliano: G. Bernardoni, E G. Brigola, Editori, 1873).

[7] Andrew Revkin, *Global Warming: Understanding the Forecast.* New York, Abbeville Press, 1992.

[8] IGBP Newsletter 41, P. Crutzen & Eugene Stoermer, 2000.

The Anthropocene: A New Epoch of Geological Time? Conference at the Geological Society of London, Burlington House, 11 May 2011, convened by Michael Ellis, Jan Zalasiewicz, Mark Williams and Alan Haywood, sponsored by the British Geological Survey. The plenary lecture was given by Nobel laureate Professor Paul Crutzen, discoverer of the 'ozone hole'.

[9] Zalasiewicz, J. *et al.* (2008). Are we now living in the Anthropocene? GSA Today vol.18 (2), p.4–8.

[10] Crutzen, P. J., and E. F. Stoermer, 2000, The 'Anthropocene'. Global Change Newsletter vol.41, p.17–18.

[11] These options are presented by Zalasiewicz *et al.*, Are we now living in the Anthropocene?, *op cit.*

[12] The Elsevier journal 'Anthropocene'.

[13] A strong advocate of the Anthropocene is Will Steffen, executive director of the Australian National University Climate Change Institute; Steffen, W., Crutzen, P.J. and McNeill J.R., 2007, The Anthropocene: are humans now overwhelming the great forces of Nature? AMBIO, vol.36, p.614–621, Royal Swedish Academy of Sciences.

[14] For a wealth of information, see Martin J.S. Rudwick, *The Great Devonian Controversy: The Shaping of Scientific Knowledge among Gentlemanly Specialists*, University of Chicago Press, 1985. See also the blog by Kelley Ross at http://www.friesian.com/rudwick.htm, last accessed 15 December 2013.

[15] John Ruskin, 1853, *The Stones of Venice*, Abridged Classic Penguin edition 2001, London, 254 pages, quote from p.13.

Chapter 6 The Great Snowball Fight
[1] The Five professors project involved Martin Brasier, a micropalaeontologist from Oxford, who joined me in delving into the deepest sedimentary rocks in Oman, whereas Albert Matter from Bern, Switzerland, John Grotzinger from MIT (now at Caltech), and Charlotte Schreiber from Queen's College, New York focused on somewhat younger rocks that were known to host valuable hydrocarbons.

[2] The global distribution of water is summarized in E.K. Berner and R.A. Berner, 1987, *The Global Water Cycle: Geochemistry and Environment*, Prentice-Hall, Englewood Cliffs, New Jersey, 397 pages.

[3] The triple junction is the point where all three phases, solid, liquid and gas co-exist.

[4] In this context, 'sublimation' refers to a solid substance changing directly to vapour, and 'deposition' refers to a vapour changing directly into a solid, like the formation of frost.

[5] Mars exploration programme, National Aeronautics and Space Administration, USA. The Mars science laboratory exploration rover 'Curiosity' was launched on 26 November 2011, and landed on 5 August 2012 in Gale Crater. It continues to explore at the time of writing, and has obtained spectacular results, http://mars.jpl.nasa.gov/msl/.

[6] K for Kelvin, or absolute temperature. K is zero (absolute zero) at -273 Celsius.

[7] Further details on the spectra for solar and terrestrial radiation can be found in textbooks on physical climatology, such as W.D. Sellers, 1965, *Physical Climatology*, University of Chicago Press, Chicago, 272 pages.

[8] Phanerozoic means the era of visible life, and extends from 543 million years ago to the present. However, life was present well before the Phanerozoic, but it was predominantly microbial. Fossils with hard skeletons or shells that are preserved as fossils occur in a great radiation of life that marks the beginning of the Phanerozoic. The oldest geological period of the Phanerozoic is the Cambrian, so the great diversification of life is known as the 'Cambrian radiation'.

[9] Cuvier's analysis of fossils led him to believe that changes were abrupt and violent, a concept known as *catastrophism*, whereas the gradual changes favoured by the Scottish geologist James Hutton were part of *uniformitarianism*.

[10] Hays, J.D., Imbrie, J. and Shackleton, N.J., 1976, Variations in the Earth's orbit: Pacemaker of the Ice Ages. Science, vo.194, (4270), p.1121-1132, doi:10.1126/science.194.4270.1121.

[11] A review is found in A.G. Dawson, 1992, *Ice Age Earth: Late Quaternary Geology and Climate*, Routledge, London, 293 pages.

[12] Nir Shaviv (Racah Institute of Physics, Jerusalem), The spiral structure of the Milky Way, cosmic rays, and ice age epochs on Earth. New Astonomer, vol.8, p.39-77 (2002).

[13] The idea is popularized in *The Chilling Stars: A Cosmic View of Climate Change*, by Henrik Svensmark and Nigel Calder, Icon Books, UK, Cambridge, 2007, 268 pages.

[14] The key papers are Shaviv, Nir J. & Veizer, Ján, 2003, Celestial driver of Phanerozoic climate change? GSA Today vol.13 (7), p.4–10, doi:10.1130/1052-5173(2003)013<0004:CDOPC>2.0.CO;2, and Veizer, J., Godderis, Y. & Francois. L.M., Evidence for decoupling of atmospheric CO_2 and global climate during the Phanerozoic eon. Nature, vol.408, p.698-701 (2000).

[15] Are there connections between the Earth's magnetic field and climate?" by V. Courtillot, Y. Gallet, J.-L. Le Mouël, F. Fluteau, A. Genevey., Earth and Planetary Science Letters, 2007, vol.253, p328.

[16] Veizer, Ján; Godderis, Yves; François, Louis M., 2000, Evidence for decoupling of atmospheric CO_2 and global climate during the Phanerozoic eon. Nature, vol.408, p.698-701 (7 December 2000), doi:10.1038/35047044.

[17] A.R. Alderman & C.E. Tilley, 1960, Douglas Mawson, 1882-1958. Biographical Memoirs of Fellows of the Royal Society vol.5, p.119–127.

[18] W.B. Harland, 1964, Critical evidence for a great infra-Cambrian glaciation. International Journal of Earth Sciences, vol. 54 (1), p.45–61.

[19] 'Infracambrian' is here used to mean 'below the Cambrian', which places the Svalbard glacial deposits in the Neoproterozoic.

[20] W.B. Harland, 2007, Origins and assessment of Snowball Earth hypotheses. Geological Magazine, vol.144, p.633-642.

[21] M.I. Budyko, 1969, Effect of solar radiation variation on climate of Earth. Tellus A vol.21 (5), p.611–1969.

[22] Kirschvink, Joseph, 1992, Late Proterozoic low-latitude global glaciation: the Snowball Earth. In: J. W. Schopf & C. Klein. *The Proterozoic Biosphere: A Multidisciplinary Study*. Cambridge University Press, p.51-52, 1348 pages.

[23] Since the lighter isotope ^{12}C is used in metabolic activity, marine organisms become enriched in ^{12}C compared to the heavier isotope ^{13}C and the water consequently becomes relatively enriched in the heavier isotope. Carbonate rocks (limestones) precipitated in these waters will have positive values of the ratio of ^{13}C/^{12}C, indicating biological productivity. If biological activity ceases, the ratio should indicate the relative abundances of the lighter and heavier isotopes of carbon supplied by volcanoes, which gives a negative value of ^{13}C/^{12}C.

[24] Hoffman, P.F., Kaufman, A.J., Halverson, G.P. & Schrag, D.P., 1998, A Neoproterozoic Snowball Earth. Science, vol.281 (5381), p.1342–1346.

[25] Hoffman, P.F. & Schrag, D.P., 2002, The Snowball Earth hypothesis: Testing the limits of global change. Terra Nova, vol.14, p.129–155.

[26] Web of Science is a database of scientific publications, created by Thomson Reuters. It serves as a research tool, but is most widely used to obtain metrics of citations to published scientific work.

[27] Ernest Rutherford claimed that all good science should be possible to explain to a barmaid. I have no intention of putting barmaids in any particular position in the pyramid of intellectual capability, especially if customers spend time discussing their latest research findings with them while drinking a pint of Greene King ale.

[28] Allen, P.A. & Etienne, J.L., 2008, Sedimentary challenge to Snowball Earth. Nature Geoscience, vol.4, p.817-825, doi:10.1038/ngeo355.

[29] Louis Agassiz, 1837, reported in E.P. Evans: The Authorship of the Glacial Theory, North American Review, vol.145, Issue 368, July 1887.

[30] Etienne, J.L., Allen, P.A., Rieu, R. & Le Guerroué, E., 2007, Neoproterozoic glaciated basins: a critical review of the Snowball Earth hypothesis by comparison with Phanerozoic basins. In: Hambrey, M.J., Christoffersen, P., Glasser, N.F. & Hubbard, B. (editors), *Glacial Sedimentary Processes and Products*, International Association of Sedimentologists Special Publication No. 39, p.343-399.

Chapter 7 Blind Dates: the Antiquity of the Earth

[1] John Lightfoot, vice Chancellor of Cambridge University, published his estimate of the age of the Earth in 1644, six years before Archbishop Ussher's statement in 1650, according to G.B. Dalrymple (1991, *The Age of the Earth*, 474 pages, Stanford University Press, Stanford, California, USA, p.14). There is some confusion as to the priority given to Lightfoot and Ussher. Lightfoot used the Bible to arrive at the date of 3929 B.C., whereas Ussher was extremely exact – 9 o'clock in the morning on October 26, 4004 B.C.

[2] John Ruskin, author of *Stones of Venice*, was a vehement opponent of Renaissance architecture and a strong advocate of Gothic. He referred to 'this pestilent art of the Renaissance'.

[3] A history of the Geology Department is found in *In Marble Halls*, by Patrick Wyse Jackson, 135 pages, published by Department of Geology, Trinity College Dublin (1994).

[4] Poem in memory of Adrian Phillips, an inspirational geologist but more importantly a wonderful, kind and generous man. Within a year of retiring from the staff at the Department of Geology, Trinity College Dublin, he was gone. He loved the Wicklow Hills, south of Dublin and was devoted to Trinity College.

[5] G. Schroeder, 1997, *The Science of God: The Convergence of Scientific and Biblical Wisdom*, p.97, The Free Press, New York, 226 pages.

[6] Steve Jones, *The Serpent's Promise*, Chapter 1, In the Beginning, *op cit.*

[7] A much fuller survey is carried out by R. Forster and P. Marston, 1999, *Reason, Science and Faith*, Monarch Publications, 479 pages.

[8] Dick Taverne, *The March of Unreason*, Oxford University Press, 2005, 310 pages.

[9] Steve Jones, page 16, *The Serpent's Promise*, *op cit.*

[10] Isaac Newton, *Philosophiæ Naturalis Principia Mathematica* (1687).

[11] John Beaumont (1650-1731) wrote a review of Dr Burnet's *Theory of the Earth*, 1693; John Ray's correspondence on John Woodward's book *An Essay Toward a Natural History of the Earth*, 1695.

[12] Benoît de Maillet, original published in French in 1748, translated in 1968. *Teliamed, or conversations between an Indian philosopher and a French missionary on the diminution of the sea.* Translated and edited by Albert V. Carozzi. University of Illinois Press, Urbana, Chicago & London.

[13] Georges-Louis Leclerc de Buffon, *Epoques de la Nature*, 1778, Garnier Frères, Paris, 670 pages.

[14] James Hutton, 1788, *Theory of the Earth*, Transactions of the Royal Society of Edinburgh, vol.1, p.217.

[15] Burchfield (1975) and Lindley (2004) cover Kelvin's role in the controversy of the age of the Earth in detail. Burchfield, J.D., 1975, *Lord Kelvin and the Age of the Earth*: New York, Science History Publications, 260 pages. Lindley, D., 2004, *Degrees Kelvin*: Washington, D.C., Joseph Henry Press, 366 pages.

[16] Curie, P. and Laborde, A., 1903, Sur la chaleur dégagée spontanément par les sels de radium, Comptes Rendus de l'Académie des Sciences, vol.136, p.673–675.

[17] Calculations by England, P., Molnar, P. & Richter, F. in 2007, John Perry's neglected critique of Kelvin's age for the Earth: A missed opportunity in geodynamics. GSA Today, vol.17 (1), p.4–9. doi:10.1130/GSAT01701A.1.

[18] O. Fisher, 1881, *The Physics of the Earth's Crust*, Murray, London, 299 pages.

[19] England et al., GSA Today, *op cit.*

[20] Richter, F.M., 1986, Kelvin and the age of the Earth: Journal of Geology, vol.94, p.395–401.

[21] Burchfield, J.D., 1975, *Lord Kelvin and the Age of the Earth*: New York, Science History Publications, p.165-166, 260 pages.

[22] The author of what I believe to be the finest textbook in geology entitled *Principles of Physical Geology*, published by Nelson. Holmes's biographer is Cherry Lewis, *The Dating Game*, Cambridge University Press, 2000, 253 pages.

[23] Patterson, C., 1956, Age of Meteorites and the Earth. Geochimica et Cosmochimica Acta, vol.10, p.230–237.

Chapter 8 Killer Blows: Fire and Brimstone

[1] Stephen Jay Gould, Simon Conway Morris, Richard Fortey, Andrew Knoll, Derek Briggs, Martin Brasier to name a few.

[2] Raup, D. & Sepkoski, J., 1982, Mass extinctions in the marine fossil record. *Science* vol.215, p.1501–1503, DOI:10.1126/science.215.4539.1501. The chart was updated in Sepkoski, J., 2002, A Compendium of Fossil Marine Animal Genera

(eds. Jablonski, D. & Foote, M.) Bulletin American Paleontology vol.363 (Paleontological Research Institution, Ithaca, New York).

[3] A highly informative account is given by A. Hallam in *Catastrophes and Lesser Calamities: The Causes of Mass Extinctions*, Oxford University Press, 2004, 226 pages.

[4] Alroy, J., 2008, Dynamics of origination and extinction in the marine fossil record. Proceedings of the National Academy of Sciences of the United States of America vol.105 (Supplement 1), p.11536–11542, doi:10.1073/pnas.0802597105. McGhee, G.R., Sheehan, P.M., Bottjer, D.J. & Droser, M.L., 2011, Ecological ranking of Phanerozoic biodiversity crises: The Serpukhovian (early Carboniferous) crisis had a greater ecological impact than the end-Ordovician. Geology, vol.40 (2), p.147, doi:10.1130/G32679.1.

[5] Table 9.1, page 162, in A. Hallam, *Catastrophes and Lesser Calamities*, Oxford University Press, *op cit.*

[6] The course of the debate over the cyclicity of extinction rate is told by Tony Hallam in *Catastrophes and Lesser Calamities*, *op cit.*, pages 156-160.

[7] Raup, D.M. & Sepkoski, J.J., 1984, Periodicity of extinctions in the geologic past. Proceedings of the National Academy of Sciences, vol.881, p.801-805.

[8] Science, vol. 221, p935, 1983; Science News, vol. 124, p.212, 1983. Hallam, *op cit.* page 160.

[9] See, for example, M.J. Benton, 1995, Diversification and extinction in the history of life, Science, vol.268, issue 5207, p.52-58. A discussion-reply exchange is found in Science, vol.269, issue 5224, p.619-620.

[10] Samuel Taylor Coleridge (1772-1834), *Rime of the Ancient Mariner*.

[11] See Derek Ager's *The New Catastrophism*, 1993, Cambridge University Press, 231 pages, or Vincent Courtillot's *Evolutionary Catastrophes: The Science of Mass Extinction*, 1999, chapter 1, Cambridge University Press, 173 pages.

[12] Alvarez, L.W., Alvarez, W., Asaro, F. & Michel, H.V., 1980, Extraterrestrial Cause for the Cretaceous-Tertiary Extinction: Experiment and Theory. *Science*, vol.208 (4448), p.1095–1108, doi:10.1126/science.208.4448.1095.

[13] Chapter 6 in *Evolutionary Catastrophes*, Courtillot, *op cit.*, p.88-100.

[14] Magnetic time zones, or chrons, have one period of normal polarity and one of reversed polarity. The main phase of eruption was in the reversed polarity part of chron 29, or C29r. The palaeomagnetic work was published in 1985.

[15] V. Coutillot, Mass extinction: seven traps and one impact? Israeli Journal of Earth Sciences, Jerusalem, vol.43, p.255-266, 1994.

[16] J. Smit, J. Hertogen, Nature vol.285, p.198 (1980). J. Smit, W. Alvarez, A. Montanari, P. Claeys, J. M. Grajales-Nishimura (1996) Special Paper Geological Society of America, vol.307, p.151.

[17] From argon dating using a laser.

[18] From Walter Alvarez, 1997, *T. Rex and the Crater of Doom*, 185 pages, Princeton University Press, Princeton, New Jersey.

[19] Chicxulub seismic Experiment, Morgan, J.V. and M.R. Warner, 1999, Chicxulub: The third dimension of a multiring impact basin. Geology, vol.27, p.407–410. Morgan, J.V. and 19 colleagues, 1997, Size and morphology of the Chicxulub impact crater. Nature, vol.390, p.472–476. Vermeesch, P.M., Morgan, J.V., 2008, Structural uplift beneath the Chicxulub impact structure, Journal Geophysical Research, vol.113, B07103, doi:10.1029.

[20] Collins G.S., Melosh H.J. & Osinski, G.R., 2012, The Impact-Cratering Process, *Elements*, vol.8, ISSN:1811-5209, p.25-30. Collins, G.S., Melosh, H.J., Morgan, J.V., *et al.*, 2002, Hydrocode Simulations of Chicxulub crater collapse and peak-ring formation, *Icarus*, vol.157, ISSN:0019-1035, p.24-33.

[21] Tony Hallam, p.6, *Catastrophes and Lesser Calamities*, attributes this phrase to T.S. Elliot (1888-1965) in *The Hollow Men* (1925).

[22] Gerta Keller, Thierry Adatte, Alfonso Pardo Juez & Jose G. Lopez-Oliva, 2009, New evidence concerning the age and biotic effects of the Chicxulub impact in NE Mexico. Journal of the Geological Society, vol.166 (3), p.393–411, doi:10.1144/0016-76492008-116. Gerta Keller, Thierry Adatte, Zsolt Berner, Markus Harting, Gerald Baum, Michael Prauss, Abdel Tantawy and Doris Stueben, 2007, Chicxulub impact predates K–T boundary: New evidence from Brazos, Texas. Earth and Planetary Science Letters, vol.255 (3–4), p.339-356, doi:10.1016/j.epsl.2006.12.026.

[23] Peter Schulte and 40 others, 2010, The Chicxulub asteroid impact and mass extinction at the Cretaceous-Paleogene boundary. Science, vol.327, p.1214-1218, doi:10.1126/science.1177265.

[24] Kuhn, T.S. *The Structure of Scientific Revolutions*. Chicago: University of Chicago Press, 1962, 264 pages, ISBN 0-226-45808-3.

[25] The phrase is tenuously attributed to British Prime Minister Bejamin Disraeli (1804-1882) but was populraized in North America by Mark Twain.

[26] V. Courtillot, *Evolutionary Catastrophes*, *op cit.*, chapter 9, p.135-143.

[27] C. Officer & J. Page, *The Great Dinosaur Extinction Controversy*, Reading MA, Helix Books, Addison-Wesley, 209 pages, 1996.

Chapter 9 An Olive Leaf and a Dove: Flood Stories

[1] A recent exploration of Noah's Flood and its impact on scientific thinking is given by David R. Montgomery in *The Rock's Don't Lie: A Geologist Investigates Noah's Flood*, W.W. Norton & Co., 320 pages, 2013.

[2] D. Ager, 1993, *The New Catastrophism*. Cambridge University Press, 231 pages. Quote is from page xix.

[3] In contrast with the geologically inspired views of James Hutton (1726-1797), who in *Theory of the Earth* referred to 'no vestige of a beginning, no prospect of an end'.

[4] S. Jaki, 1974, *Science and Creation*. Scottish Academic Press, 367 pages.

[5] The calculations of the amount of water in the hydrosphere available to flood the entire world are my own, but in the 17th century a Dutch theologian named Isaac Vossius (1618-1689) came to the same conclusion, and thereby argued that the Flood was local, not global. The Anglican Bishop, Edward Stillingfleet (1635-1699), took the same view in 1666. There were several ways out of this problem: (1) to invoke the existence of a watery canopy above the Earth, which drained downwards to supply Noah's flood, as suggested by Edmund Halley, (2) the eruption of a primordial ocean below the surface of the Earth (Thomas Burnet, 1681, in *Sacred Theory of the Earth*), and (3) the capturing of water from a passing comet, suggested by William Whiston. All three of these imaginative explanations left the intractable problem of where all this water went *after* the Flood - a problem

recognized in the 17th century by Leonardo. Burnet believed that it drained back through cracks in the floor of the oceans.

[6] This is very thoroughly set out in Richard Huggett's book *Cataclysms and Earth History*, 1989, Clarendon Press, Oxford, 220 pages.

[7] C.C. Gillespie, 1951, *Genesis and Geology*, Harper and Row, New York, 315 pages.

[8] Georges Cuvier, *Discours sur les révolutions de la surface du globe*. Issued as a preface to *Récherches sur les ossemens fossiles* (1812).

[9] G. L. Herries Davies, 1969, *The Earth in Decay: A History of British Geomorphology 1578-1878*, p.251, MacDonald, London, 390 pages.

[10] N. Cohn, 1996, *Noah's Flood: The Genesis Story in Western Thought*, p 121, Yale University Press, New Haven, 154 pages.

[11] Arthur Schopenhauer wrote 'Talent hits a target no-one else can hit; genius hits a target no-one else can see'.

[12] Georges-Louis Leclerc, Comte de Buffon, Les *Epoques de la Nature*, 1778, De l'Imprimerie royale, 264 pages.

[13] Georg Kirchmaier, 1667, *De diluvii universalitate dissertatio prolusoria*. Geneva, p.3-60.

[14] Quotes from a letter written by Bretz in later years, looking back at the controversy (1978).

[15] Baker V.R. 1973, Paleohydrology and sedimentology of Lake Missoula flooding in eastern Washington. Geological Society of America Special Paper 144, p.1-79.

[16] Baker, V.R., 1996, Megafloods and glaciation. In: *Global Changes in Post Glacial Times: Quaternary and Permo-Carboniferous* (ed. by I.P. Martini), Oxford University Press, Oxford, p.98-108, 355 pages.

[17] Smith, A. J., 1985, A catastrophic origin for the paleovalley system of the eastern English-Channel. Marine Geology, vol.64, p.65–75. Roep, T. B., Holst, H., Vissers, R. L. M., Pagnier, H. & Postma, D., 1975, Deposits of southward-flowing, Pleistocene rivers in the Channel region, near Wissant, N.W. France. Palaeogeography Palaeoclimatology, Palaeoecology, vol.17, p.289–308.

[18] S. Gupta, J.S. Collier, A. Palmer-Felgate & G. Potter, 2007, Catastrophic flooding origin of shelf valley systems in the English Channel. Nature, vol.448, p.342-345.

[19] Gibbard, P. L., 1995, in *Island Britain: a Quaternary Perspective* (ed. Preece, R. C.) p.15–26, Geological Society Special Publication, London.

[20] Preece, R. C. *Island Britain: a Quaternary Perspective*, 1995, *op cit.*

[21] Warren, J.K., 2006, *Evaporites: sediments, resources and hydrocarbons*. Birkhäuser, 352 pages, ISBN 978-3-540-26011-0.

[22] W.B.F. Ryan, 2009, Decoding the Mediterranean salinity crisis. Sedimentology, vol.56, p.95-136.

[23] W. Ryan and W. Pitman, 1998, *Noah's Flood: The New Scientific Discoveries about the Event that Changed History*. Simon & Schuster, New York 319 pages.

[24] A press release from the US National Geographic Society produced a flurry of reports in national newspapers, including *The Irish Times*, Saturday 16 September 2000.

[25] UNESCO and International Union of Geological Sciences, Project 521, http://sealevel.ca/IGCP521/.

[26] A 182-page special issue of the journal *Quaternary International* was published in 2012 resulting from this multidisciplinary project. Quaternary International, vol.261, p.1-182 (30 May 2012), Edited by Valentina Yanko-Hombach, Nicolae Panin, Mariana Filipova-Marinova and Norm Catto, *IGCP 521: Caspian-Black Sea – Mediterranean Corridors during the Last 30 ka: Sea Level Change and Human Adaptive Strategies, Selected Papers.* Extended abstracts from plenary meetings related to this IGCP 521 project are also available.

[27] This view is held by Ukrainian and Russian scientists, particularly Valentina Yanko-Hombach. Yanko-Hombach, Valentina, 2007, *The Black Sea Flood Question: Changes in Coastline, Climate and Human Settlement.* Springer, 971 pages ISBN 1-4020-4774-6.

[28] Giosan, Liviu *et al.*, 2009, Was the Black Sea catastrophically flooded in the early Holocene? Quaternary Science Reviews, vol.28 (12-2), p.1-6, doi:10.1016/j.quascirev.2008.10.012.

[29] R. Ward, 1978, *Floods: A Geographical Perspective.* London, Macmillan, 244 pages.

[30] 111b, Plato. *Plato in Twelve Volumes*, Vol. 9 translated by W.R.M. Lamb. Cambridge, MA, Harvard University Press; London, William Heinemann Ltd., 1925.

[31] Steve Jones writes (p.244) in *The Serpent's Promise*, 'The story, with its ten plagues . . . . [which] is among the most familiar in scripture . . . is not matched by any evidence that the events recorded ever took place'. This view echoes that of some Egyptologists and archaeologists, who deny that there was an Exodus, and instead believe that the Biblical narrative is a confused account of the expulsion of the Hyksos people (Canaanites from southwestern Asia), written several (7 to 8) centuries later, recorded in the Torah.

[32] *Exodus* chapter 7, verse 1, to chapter 12, verse 30, *Holy Bible.*

[33] For greater detail, see Barbara J. Sivertsen's *The Parting of the Sea.* Princeton University Press, New Jersey, 239 pages, 2009, and Ian Wilson's *Exodus: the True Story*, 1985, Harper & Row Publishers, 208 pages. Stephen B Segall writes in *Understanding the Exodus and Other Mysteries of Jewish History* (218 pages, Etz Haim Press, Ann Arbor, Michigan, 2003), p.99, 'Most of the miracles described in the story of the Exodus appear to have some basis in fact. However, to the extent that these phenomena did occur they were natural events that were interpreted as being supernatural by the people who witnessed them'.

[34] Some support for this amalgamation comes from Psalm 105, which only mentions 7 plagues – darkness, which appears out of order at the beginning of the section in Psalm 105 between verses 28 and 36, water into blood, frogs, flies and gnats which seem to be combined, hail, locusts and finally the plague on the firstborn. The plague on livestock and the plague of boils are omitted. The Psalms are clearly structured along literary, poetic or musical lines, and the reduction of plagues to seven and the slight re-ordering simply seems to be a convenient way of recalling the events leading to the Exodus. The books of Genesis to Numbers appear to have three sources, and some mixing of plagues may have taken place by amalgamation of material from these different sources.

[35] The Sumerian goddess Inanna and the Egyptian text *Admonitions of Ipwer* both refer to turning water to blood. From M. Lichtheim, *Ancient Egyptian Literature, vol 1, The Old and Middle Kingdoms*, University of California Press, Berkeley, 1973, p.151.

[36] M.A. Summerfield & N.J. Hulton, 1994, Natural controls on fluvial denudation rates in major world drainage basins. Journal of Geophysical Research, vol.99, p.13871-13883.

[37] Ian Wilson, *Exodus: the True Story*, 1985, Harper Row Publishers, p.113, 208 pages.

[38] Two compilations are *The Archaeology of Geological Catastrophes*, edited by W.J. McGuire, D.R. Griffiths, P.L. Hancock and I.S. Stewart, 2000, Geological Society Special Publication 171, and *Santorini Volcano*, edited by T,H. Druitt and 7 others, 1999, Geological Society Memoir 19.

[39] J.V. Luce, 1969, *The End of Atlantis*. Paladin, 187 pages.

[40] The excavations of Akrotiri by Marianatos in the 1960's found a city buried by pumice.

[41] G. Galanopoulos, 1971, The Eastern Mediterranean trilogy in the Bronze Age. In: Kaloueropoulou, A. (ed.) *Acta of the First International Scientific Congress on the Volcano of Thera*, Athens, p.184-210.

[42] See, for instance, Barbara Sivertsen's *The Parting of the Sea*, op cit.

[43] W.L. Friedrich, P. Wagner & H. Tauber, 1990, Radiocarbon dated plant remains from the Akrotiri excavation on Santorini, Greece. In: Hardy, D.A. (ed.) *Thera and the Aegean World III*, 3. The Thera Foundation, London, p.188-196.

[44] S. Manning, 1988, The Bronze Age eruption of Thera: absolute dating, Aegean chronology and Mediterranean cultural interrelations. Journal of Mediterranean Archaeology vol.1, p.17-82.

[45] H.H. Rowley, 1950, *From Joseph to Joshua: Biblical Traditions in the Light of Archaeology*, Oxford University Press, 214 pages.

[46] H.H. Rowley, 1950, *From Joseph to Joshua: Biblical Traditions in the Light of Archaeology*, op cit.

[47] Dates range between 1900 and 1350 BCE. One way of estimating the time period of Joseph in Egypt is to work back from the Exodus, at 1447 BCE. We therefore need to know the length of the Israelite sojourn in Egypt. Translations from the Hebrew (so-called Masoretic) text suggest 430 years, but Josephus, who lived in the first century AD, and whose writings therefore pre-date the compilation of the Hebrew Old Testament (4th century AD) by several hundred years, stated in his *Antiquities of the Jews* (Ch. XV:2) that the Israelites left Egypt ….'430 years after our forefather Abraham came into Canaan, but 215 years only after [Joseph's father] Jacob removed into Egypt'. The Greek version of the Old Testament (Septuagint), the oldest version of which is also older than the Masoretic text, states that the period ' . . . in which they sojourned in the land of Egypt *and in the land of Canaan* was 430 years'. So it looks as if something was dropped out of the Hebrew text in the intervening period between the writing of the Septuagint and the Masoretic version from which the translation in the Bible is derived - an error or omission introduced in successive translations.

[48] I am greatly indebted to the fine detective work of David Roth in his book *A Test of Time: The Bible – from Myth to History*, 1995, Arrow, 425 pages.

[49] C.J. Humphreys & R.S. White, 1995, The eruption of Santorini and the date and historicity of Joseph. Science and Christian Belief, vol.7, p.151-162. The Hyksos pharaohs were foreigners, probably from Asia, who were hated by their Egyptian successors. Very little is recorded of their lives and times. Humphreys and

White suggest that an Egyptian pharaoh was less likely to appoint a Hebrew such as Joseph as a second-in-command than a foreign pharaoh. It seemed reasonable, therefore, that the Hyksos 15th Dynasty corresponded with the time of Joseph in Egypt. The 15th Dynasty is part of the Second Intermediate Period and has been conventionally given dates of 1633-1525 BC. It is known that the Hyksos kings established their stronghold at Aravis, in the eastern part of the delta region. This is the same as Rameses in Goshen, which is thought to be where Joseph settled.

[50] David Rohl, *A Test of Time: The Bible – From Myth to History*, Arrow Books Limited, 1996, London, 425 pages. According to Rohl, Joseph would have been made Vizier during the long reign of Amenemhat III, and continued in this prestigious position during the reigns of the first rulers of the 13th Dynasty.

[51] S. Thorarinsson, 1969, The Lakagigar eruption of 1783. Bulletin of Volcanology, vol.33, p.910-927; Hammer, C.V., 1977, Post volcanism revealed by Greenland ice sheet impurities. Nature, vol.270, p.482-486.

[52] Th. Thordarson & S. Self, 1993, The Laki (Shaftar Fires) and Grimsvötn eruptions in 1783-1785. Bulletin of Volcanology, vol.55, p.233-263.

Chapter 10 Deep Trouble: the Witch's Cauldron

[1] 'Dynamic Topography: A key surface record of deep Earth processes', convened by Roderick Brown, Patience Cowie, Stewart Fishwick, Gregory Houseman, Michael Kendall and Nicky White. Held 1-2 September 2011, the Geological Society, Burlington House, Piccadilly, London.

[2] Global Geodetic Observing System, and the Global Strain Rate Model, Kreemer, C., Holt, W.E. & Haines, A.J., 2003, An integrated global model of present-day plate motions and plate boundary deformations. Geophysical Journal International, vol.154, p.8-34.

[3] For a summary, see Nolet, G., 2008, *A Breviary of Seismic Tomography*, Cambridge University Press, 360 pages.

[4] See website by Gillian Foulger, www.mantleplumes.org.

[5] Don L. Anderson & Warren B. Hamilton, Zombie Science and Geoscience, www.mantleplumes.org, viewed 4/11/2013, last updated 11-12-2008.

[6] This definition more or less follows Richards, M.A. & Hager, B.H., 1984, Geoid anomalies in a dynamic Earth. Journal Geophysical Research, vol.89, p.5487-6002, though the term was coined earlier.

[7] Fred Astair sang a song written by George and Ira Gerschwin in 1937 called 'Let's call the whole thing off'. Differences in dialect between English and American have always taken on greater importance than questions of geodynamics nomenclature.

[8] Currents set up in the ocean are deflected by the effects of the Earth's rotation, causing the surface of the ocean to be tilted. The Gulf Stream is an example. This is a true 'oceanographic dynamic topography' since the inclination of the sea surface is strongly affected by the velocity of the current.

[9] Hartley, R.A., Roberts, G.G., White, N.J. & Richardson, C., 2011, Transient convective uplift of an ancient buried landscape. Nature Geoscience, vol.4, p.562-565.

[10] Hartley, R.W. & Allen, P.A., 1994, Interior cratonic basins of Africa: relation to continental break-up and role of mantle convection. Basin Research, vol.6, p.95-113.

[11] R. Moucha, A. M. Forte, J.X. Mitrovica, D.B. Rowley, S. Quéré, N.A. Simmons & S.P. Grand, 2008, Dynamic topography and long-term sea level variations: there is no such thing as a stable continental platform. Earth and Planetary Science Letters, vol.271, p.101-108.

[12] H. Poore, N. White & J. Maclennan, 2011, Ocean circulation and mantle melting controlled by radial flow of hot pulses in the Iceland plume. Nature Geoscience, vol.4, p.558-561.

[13] J.F. Rudge, M.E.S. Champion, N. White and others, 2008, A plume model of diachronous transient uplift at the Earth's surface. Earth and Planetary Science Letters, vol.267, p.146-160.

[14] White, N. & Lovell, B., 1997, Measuring the pulse of a plume with the sedimentary record. Nature, vol.387, p.888-891.

[15] Vincent Courtillot, *Evolutionary Catastrophes: The Science of Mass Extinction*, 1999, Cambridge University Press, 173 pages.

[16] Morgan, W. J., 1972, Deep mantle convection plumes and plate motions. Bulletin American Association of Petroleum Geologists, vol.56, p.203–213.

[17] This term was coined in 1992 by Coffin, M. & Eldholm, O., 1992, Volcanism and continental break-up: a global compilation of large igneous provinces. In: Storey, B.C., Alabaster, T. & Pankhurst, R.J. (eds.) *Magmatism and the Causes of Continental Breakup*. Special Publication Geological Society, London, no.68, p.17–30.

[18] Storey, B.C., 1995, The role of mantle plumes in continental breakup: Case histories from Gondwana. Nature, vol.377, p.301–308, doi:10.1038/377301a0.

[19] This view underlies the very readable *Dynamic Earth: Plates, Plumes and Mantle Convection*, by Geoffrey F. Davies, Cambridge University Press, 458 pages, 1999.

[20] G. Foulger, *Plates versus Plumes: A Geological Controversy*, Wiley Blackwell, 2010, 364 pages.

Chapter 11 Salami Slicing Earth's Outer Shell

[1] Benjamin Britten wrote four 'The Sea' interludes in the opera *Peter Grimes*, about an Aldeburgh fisherman who was hounded to his death by hostile townspeople. In one of the interludes, Peter Grimes sings 'What harbour shelters peace, away from tidal waves, away from storms?' The disappearance of this part of the Suffolk and Norfolk coasts by tide and wave provides a longer-term backdrop for the personal tragedy of Peter Grimes.

[2] From the name of Simon Winchester's book, *The Map that Changed the World: William Smith and the Birth of Modern Geology*, 2001, Harper-Collins, New York, 368 pages.

[3] Berry, W., 1987, *Growth of a Pre-Historic Time Scale based on Organic Evolution*, Oxford, Blackwell Science, p.51-59, 202 pages.

[4] An identical marble bust stands in the University Museum in Oxford.

[5] 'Red noise' refers to apparently random fluctuations, also known as Brownian motion, which in this case describes the background character of the preservation of sediments in stratigraphy.

[6] 'Shredded signals' refers to the way in which natural systems smooth out extreme events. The record of these events in sediments is therefore difficult to decipher.

[7] Vaughan, S., Bailey, R.J. & D.G. Smith, 2011, Detecting cycles in stratigraphic data: spectral analysis in the presence of red noise. Palaeoceanography vol.26, PA4211.

[8] Vail, P.R., 1992, The evolution of seismic stratigraphy and the global sea-level curve. In: Dott, R.H. Jr., Editor, *Eustasy: The Historical Ups and Downs of a Major Geological Concept*, Geological Society of America, Memoir, vol.180, p.83–91.

[9] P.A. Allen & J.R. Allen, 2013, *Basin Analysis: Principles and Application to Petroleum Play Assessment*, 3rd edition, Wiley-Blackwell, Oxford, 619 pages.

[10] Vail, P.R., Mitchum, R.M. Jr. & Thompson, S., 1977a, Relative changes of sea level from coastal onlap, p.63-82, and 1977b, Seismic stratigraphy and global changes of sea level, Part 4, Global cycles of relative changes of sea level, p.83-97 in *Seismic Stratigraphy: Applications to Hydrocarbon Exploration* (edited by C.E. Payton), American Association of Petroleum Geologists Memoir 26, and Haq B.U., Hardenbol, J. & Vail, P.R., 1988, Chronology of fluctuating sea levels since the Triassic (250 Myr ago to present), Science, vol.235, p.1156-1167.

[11] Robert M. Carter, 1998, Two models: global sea-level change and sequence stratigraphic architecture. Sedimentary Geology, vol.22, p.23-36.

[12] Vail et al., 1977, p.96, *op cit.*

[13] A.D. Miall & C.E. Miall, 2001, Sequence stratigraphy as a scientific enterprise: the evolution and persistence of conflicting paradigms. Earth Science Reviews, vol.54, p.321-348.

[14] Isaac Newton wrote this, and the words are used on the edge of the British 2 pound coin, but there are many more ancient references.

[15] Barrell, J., 1917, Rhythms and the Measurements of Geologic Time: Bulletin of the Geological Society of America, vol.28, p.745-904.

[16] Peter M. Sadler, 1981, Sediment accumulation rates and the completeness of stratigraphic successions. Journal of Geology, vol.89, p.569-584, quote is from p. 569.

[17] Larry Sloss, 1963, Sequences in the cratonic interior of North America. Geological Society of America Bulletin, vol.74, p.93-114.

[18] Derek Ager, *The New Catastrophism*, 1993, Cambridge University Press, 231 pages.

[19] Charles Darwin, 1859, *The Origin of Species by Natural Selection, or the preservation of favoured races in the struggle for life*, London, John Murray. Quote comes from p.310.

[20] Dewey, J.F. & Pitman, W.C., 1988, Sea-level changes: mechanisms, magnitudes and rates. In: J.L. Pindell & C.L. Drake (editors) *Paleogeographic Evolution and Non-Glacial Eustasy, Northern South America*, Society of Economic Paleontologists and Mineralogists, Special Publication no.58, p.1-16.

[21] Originally a short novel by Robert Louis Stevenson first published in 1886, and an American horror film from 1931. It is a story of a mild-mannered scientist who takes a potion that turns him into a homicidal maniac.

[22] C.E. Miall & A.D. Miall, 2002, The Exxon factor: the roles of corporate and academic science in the emergence and legitimation of a new global model of sequence stratigraphy. The Sociological Quarterly, vol.43/2, p.307-334.

[23] The idea of social worlds derives from Adele Clarke's (1991) analysis of the work of Anselm Strauss, which is discussed in relation to the sequence stratigraphy paradigm by Miall & Miall, 2001, *op cit.* Clarke, A., 1991, Social worlds/arenas theory as organizational theory, p.119-158, in *Social Organization and Social Progress: Essays in Honour of Anslem Strauss*, ed by D. Maines, New York, Aldine de Gruyter.
[24] A.D. Miall & C.E. Miall, 2001, p.326, *op cit.* and C.E. Miall & A.D. Miall, 2002, p.315, *op cit.*
[25] Clarke, A. & Gerson, E., 1990, Symbolic interactionism in social studies of science, in Becker, H. & McCall, M., editors, *Symbolic Interaction and Cultural Studies*, University of Chicago Press, p.179-214, cited by Miall & Miall p.327, A.D. Miall & C.E. Miall, 2001, Sequence stratigraphy as a scientific enterprise: the evolution and persistence of conflicting paradigms. Earth Science Reviews, vol.54, p.321-348.
[26] J.R. Underhill, 1991, Controls on Late Jurassic seismic sequences, Inner Moray Firth, UK North Sea: A critical test of a key segment of Exxon's original global cycle chart. Basin Research, vol.3, p.79-98.
[27] Spasojevic, S., Liu, K., Gurnis, M. & R.D. Müller, 2008, The case for dynamic subsidence of the US east coast since the Eocene. Geophysical Research Letters, vol.35, L08305.
[28] For example, NEFTEX, www.neftex.com.
[29] John C. Tipper, 1994, Some tests of the Exxon Global Cycle Chart. Mathematical Geology, vol.26, p.843-855.
[30] Andrew D Miall, 1992, Exxon global cycle chart: An event for every occasion? Geology, vol.20, p.787-790.

Chapter 12 Public Enemy Number 1: Carbon Crisis?

[1] Davis D.L. & Bates D., A Look Back at the London Smog of 1952 and the Half Century Since; A Half Century Later: Recollections of the London Fog (Environmental Health Perspectives, December 2002).
[2] Jeremy Leggett, *The Carbon War; Dispatches from the End of the Oil Century*, Allen Lane, London, 1999, 352 pages.
[3] Bryan Lovell, *Challenged by Carbon; The Oil Industry and Climate Change*, Cambridge University Press, 212 pages, 2010.
[4] Raven, J. A. & P. G. Falkowski, 1999, Oceanic sinks for atmospheric CO_2. Plant, Cell and Environment, vol.22, p.741-755.
[5] Cox, P.M., Betts, R.A., Jones, C.D., Spall, S.A. & Totterdell, I.J., 2000, Acceleration of global warming due to carbon-cycle feedbacks in a coupled climate model. Nature, vol.408, p.184-187.
[6] Particularly informative material on this slow-down in the rate of warming is provided by the UK Meteorological Office (http://www.metoffice.gov.uk/research/news/recent-pause-in-warming) and the Yale Climate Project (http://www.yaleclimatemediaforum.org/2013/09/examining-the-recent-slow-down-in-global-warming).
[7] Bryan Lovell, former President of the Geological Society (2010-2013), in *Challenged by Carbon*, 2010, *op cit.*

[8] *Climate Change in the Texan Mind*, Research report, Yale Climate Project, 23 September 2013. *How Americans Communicate about Global Warming*, April 2013, Research report, Yale Climate Project, 20 August 2013.

[9] S. Pacala & R. Socolow, 2004, Stabilization wedges: solving the climate problem for the next 50 years with current technologies, Science, vol.305, p.968-972. Part of the *Carbon Mitigation Initiative*, Princeton University, funded by Ford and BP.

[10] A.D. Miall & C.E. Miall wrote in 2009, 'There is ample space for the earth-science community to add its informed voice to debates about energy and climate change, but, to date, this voice appears to be have been largely ineffective. Geoscience Canada, vol.36/1, p.33-41, The Geoscience of Energy and Climate 1. Understanding the climate system, and the Consequences of Climate Change for the Exploitation and Management of Natural resources: the View from Banff.

[11] The United States Senate Committee on Energy & Natural Resources held a hearing in 2012 entitled, "Induced Seismicity Potential in Energy Technologies." Dr. Murray Hitzman, the Charles F. Fogarty Professor of Economic Geology in the Department of Geology and Geological Engineering at the Colorado School of Mines in Golden testified.

[12] For an overview of the role of geology in choosing storage sites, see Kaldi, J.G, Gibson-Poole, C.M. & Payenberg, T.H.D., 2009, Geological input to selection and evaluation of CO_2 geosequestration sites, in Grobe, M., Pashin, J.C. & Dodge, R.L., editors, *Carbon Dioxide Sequestration in Geological Media – State of the Science*, American Association of Petroleum Geologists Studies in Geology no.59, p.5-16.

[13] Q. Schiermeier, 2006, p. 621, Putting the carbon back: the hundred billion tonne challenge. Nature, vol.442, p.620-623.

[14] Chadwick , A., Arts, R., Bernstone, C., May, F., Thibeau, S. & Zweigel, P., 2008, *Best Practice for the Storage of CO_2 in Saline Aquifers*. British Geological Survey, Keyworth, Nottingham.

[15] Kelemen, P.B. & Matter, J., 2008, *In situ* carbonation of peridotite for CO_2 storage. Proceedings of the National Academy of Sciences of the USA, vol.105, p.19295-17300.

Chapter 13 Epilogue: Dark Sarcasm in the Classroom

[1] American philosopher Richard Rorty (1931-2007). Rorty also rejected the view that science can depict the world, though he was criticized for it.

[2] From Pink Floyd, The Wall, 1979.

[3] Nahmanides, *Commentary on the Torah*. Edited by C. Chavel, Jerusalem, Rav Kook Institute, 1971 (English translation).

[4] J.C. Polkinghorne, 1998, *Belief in God in an Age of Science*, Yale University Press, New Haven and London, 133 pages.

[5] J.C. Polkinghorne, 1989, *Science and Providence: God's Interaction with the World*, SPCK, 144 pages, chapter 5.

[6] Bill McKibben, 1990, *The End of Nature*. Anchor Books, New York, 226 pages.

[7] McGrath, A.E. 1998, *The Foundations of Dialogue in Science and Religion*. Blackwell, Oxford, 256 pages, p.208. McGrath illustrates the difference between natural science and religion by quoting the English literary critic and author C.S. Lewis, who developed these ideas in a sermon in 1941.

[8] Jacob Bronowski (p.20), The Abacus and the Rose, Science and Human Values, 1972, Harper & Row, 119 pages.

ABOUT THE AUTHOR

Philip Allen was born in Bath, Somerset in 1953, and graduated in geology from the University of Wales, Aberystwyth in 1974. After two years in the oil industry he studied for a doctorate at Cambridge University supervised by Dr Peter Friend, then moved to a postdoctoral position in the University of Bern, Switzerland where he worked with Professor Albert Matter. He held lectureships in Cardiff University and Oxford University before taking the Chair of Geology and Mineralogy at Trinity College Dublin, followed by professorships at ETH Zürich and Imperial College London. He is the author of three editions of *Basin Analysis: Principles and Applications* (1990, 2005 and 2013) with his twin brother John, and *Earth Surface Processes* published in 1997. He has published over 100 scientific papers in the fields of sedimentology, stratigraphy, basin analysis and Earth surface processes.

Philip Allen served as Senior Proctor of Oxford University in 2004. He was given a Royal Society-Wolfson research merit award for 2006-2011 held at Imperial College, and the Lyell Medal of the Geological Society of London in 2007. He served as Science secretary of the Geological Society 2009-2012.

www.ingramcontent.com/pod-product-compliance
Lightning Source LLC
Chambersburg PA
CBHW051637170526
45167CB00001B/226